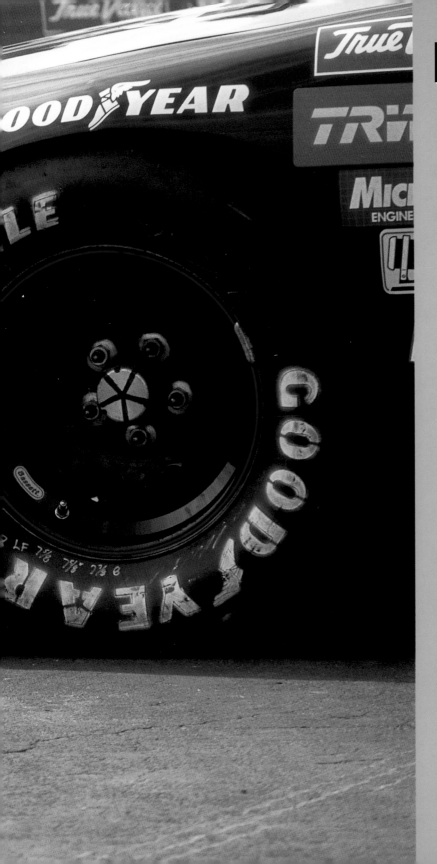

FORD, LINCOLN & MERCURY
STOCK CARS

Dr. John Craft

MBI Publishing Company

First published in 1999 by MBI Publishing Company, 729 Prospect Avenue, PO Box 1, Osceola, WI 54020-0001 USA

MBI Publishing Company books are also available at discounts in bulk quantity for industrial or sales-promotional use. For details write to Special Sales Manager at Motorbooks International Wholesalers & Distributors, 729 Prospect Avenue, Osceola, WI 54020-0001 USA.

Library of Congress Cataloging-in-Publication Data
Craft, John Albert.
    Ford, Lincoln & Mercury stock cars/John Craft.
        p. cm.—(Enthusiast color series)
    Includes index.
    ISBN 0-7603-0487-4 (pbk.: alk. paper)
    1. Stock cars—United States. 2. Ford automobile.
3. Lincoln automobile. 4. Mercury automobile. 5. Stock car racing—United States. 6. NASCAR (Association)
7. n-us. I. Title. II. Title: Ford, Lincoln, and Mercury stock cars. III. Series.
TL236.28.C23   1998
629.228—dc21   98-49598

On the front cover: Bill Elliott put brother Ernie's horsepower-making prowess to good work in 1983 in a fleet of rounded-off new-body-style Thunderbirds (an example seen here during a pit stop). From 1983 to 1986 (the life of the new body style), that duo conspired to produce 17 Winston Cup wins. *Mike Slade*

On the frontispiece: Ford race cars lost all connection to factory production cars by 1976. Fabricated panels, purpose-built bucket seats, and a jungle gym's worth of tubing characterized the control cabins of cars like David Pearson's 1976 Mercury.

On the title page: These Thunderbirds, like all NASCAR race cars, are a promotional heaven for ad agencies—especially the front end, where about half the stickers go. Teams have a deal with each company featured, where the stickers go, and how much money the company gives to the series. Some of the small stickers, like those featuring spark plug companies, are optional; the team chooses which spark plug they want to use and the company supplies the spark plugs in exchange for a sticker on the car.

On the back cover: Donnie Allison drove long-nosed Fords for legendary mechanic Banjo Matthews in 1969 and 1970. His poppy-red Talladegas were often the fastest cars on the track. A replica of one of his #27 cars is currently on display at the International Motorsports Hall of Fame in Talladega.

Edited by Tracy Snyder

Designed by Rebecca Allen

Printed in Hong Kong

Contents

Davey Allison and Bill Elliott were the fastest men in NASCAR in 1987. Both drivers regularly qualified their Thunderbirds at speeds in excess of the double ton, and between them they won eight races.

Introduction
First On Race Day

It's been said that Ford stands for First On Race Day. Without a doubt, few automobile manufacturers—either foreign or domestic—can match the Ford Motor Company's motorsports accomplishments. Indy, LeMans, F-1, NHR—Ford drivers have been there and done that—many times. Ford's dominance in four-wheeled competition has vexed rivals from "Il Commendatore" Enzo Ferrari to Bill "Grumpy" Jenkins, and the Blue Oval has graced world champions in just about every form of organized competition.

Nowhere has Ford (and sister division Lincoln Mercury) motorsports muscle been more strongly felt than in the stock car ranks. Though not major players during the first few seasons of Big Bill France's Grand National division, Fomoco-badged competition cars soon began to make their mark. By the middle of the first full decade of stock car competition, factory-backed Fomoco teams had so thoroughly flummoxed their General Motors rival that GM execs opted to slow the Blue Oval juggernaut through behind-the-scenes intrigue rather than on-the-track confrontation.

After being sidelined for a handful of seasons, Ford and Mercury teams returned to the fray and soon regained their preeminent position. Ralph Moody and John Holman served as the focus of Fomoco's 1960s racing endeavors, and their collective contributions shaped both the sport then and the series today. Competition cars and racing R&D that came out of the Holman & Moody shop led directly to the success enjoyed by such legendary drivers and mechanics as David Pearson, Robert Yates, Cale Yarborough, LeeRoy Yarbrough, Junior Johnson, Bud Moore, Fred Lorenzen, Ned Jarrett, Banjo Matthews, Leonard and Glenn Wood, Fireball Roberts, Joe Weatherly, and a host of others.

Fomoco drivers made the 427 FE, Boss 429, and 351 C engines both famous and feared in stock car circles, and the corporation's early emphasis on aerodynamics led to the focus on low coefficient of drag numbers that still holds sway today both on and off the track.

After a bleak decade-long absence from the series during the "Fuelish" 1970s, Ford racers reclaimed their places at the head of the pack and have remained there ever since "Awesome Bill" Elliott broke Bow Tie fans' hearts during the early 1980s.

As the dawn of the next century approaches, Ford is still a strong force in the NASCAR series. To date, Ford and Mercury drivers have scored more Grand National/Winston Cup wins than any other automotive manufacturer, and it's doubtful that things will be any different in the foreseeable future.

Though late 1940s, Lincolns were anything but svelte, they actually made fairly respectable race cars. It was a Lincoln, after all, that Jim Roper piloted to victory in the very first NASCAR race. Tim Flock (shown here on the Daytona Beach course) and a number of other drivers also drove hot-rod Lincolns during the first few seasons of what later came to be called Grand National racing. *Daytona Racing Archives*

1

NASCAR's First Years Through the AMA Factory Ban

NASCAR's Beginnings: Pre-1950

It's not exactly clear what Jim Roper expected that day in June of 1949 when he pulled his 1949 Lincoln up to the starting line. But it's pretty much a certainty that he had no idea what he was getting the Ford Motor Company (Fomoco) into. After all, as far as he knew, that day's race at the Charlotte Speedway was just one more of the increasingly popular sedan races that were beginning to be held across the Southland during the late 1940s. And though a charismatic giant of a fellow named Bill France was behind this particular bull-ring affair, there was really nothing to distinguish Big Bill's newly-formed National Association for Stock Car Automobile Racing (NASCAR) from the other fledgling sanctioning organizations of the day— at least at the time, that is.

The race in question was the first ever organized by NASCAR for its new "Strictly Stock Division," and Roper, along with 33 other drivers, was on hand for the 200-lap affair that Sunday afternoon in North Carolina. When the starter gave the command, Roper floored his Mecklenburg Motors-backed Lincoln with all the rest.

Like many of his competitors in the field, Roper had driven his "race" car to the track that day. Making that feat even more remarkable today is the fact that he'd driven the same car more than a thousand miles cross-country to the event (after having read about the race in a newspaper comic strip of all places!). It seems that purpose-built race cars and block-long team transporters were still more than a few years in the future back in 1949.

But that didn't mean that the on-track competition was in any way deficient. Future stars

1949-1960

like Frank "Rebel" Mundy, Fonty Flock, Lee Petty, Buck Baker, Tim Flock, Red Byron, Herb Thomas, and Curtis Turner took the green flag at the same time that Roper did, and their collective presence in the field was a virtual guarantee that the race action would be fast and furious. Lee Petty totaled his Buick in the first ever NASCAR wreck (in a car that he borrowed from a neighbor for the event);

a number of other drivers also came to grief while challenging for the lead.

When the last circuit around that dusty, rutted dirt track was complete, Glenn Dunnaway had brought his 1947 Ford home three full laps ahead of Roper's second-place Lincoln. But a post-race inspection of the winning car turned up a set of non-stock leaf springs. As a result, Roper was

Ford executives paid Bill France's NASCAR series little attention until the mid-1950s. It was then that folks in the Dearborn "Glass House" finally decided to get into factory-backed stock car competition in a big way. The first step in that process was securing the services of Indy 500 ace Pete DePaolo (right) to head up the effort. One of DePaolo's first acts was to sign up promising young driver Ralph Moody. *Daytona Racing Archives*

When Pete DePaolo stepped aside as the leader of Ford's factory stock car effort, Ralph Moody (center, checkered shirt) and John Holman (in car) stepped in to fill the void. Holman & Moody are joined here by early team drivers (from left) Curtis Turner, Joe Weatherly, Marvin Panch, and Bill Amick. *Daytona Racing Archives*

credited with the win as twilight settled over the track. First place paid a handsome $2,000 for NASCAR race number one, and Roper used some of the purse to buy a new refrigerator. As things turned out, it was the lanky Kansan's first and last NASCAR triumph. But it was not to be the last time that a Fomoco-powered race car visited a Grand National (later to become Winston Cup) winner's circle—not by a long shot.

The cars that Henry Ford built became associated with racing and organized competition right from the very start of the Ford Motor Company's operations. In fact, Ol' Henry built his first competition car as far back as 1901, when the company was but little removed from the basement shop where the very first Fomoco car (the Quadracycle) had been built in 1896. The now famous #999 Ford racer made its initial appearance in 1904, and H. F. hired legendary driver Barney Oldfield to pilot it to a number of speed records. That particular Ford was the first automobile to exceed 90 miles per hour—a phenomenal velocity for its day.

It didn't take long for Ford to begin to appreciate the positive effect on sales that racing headlines produced.

Ford's corporate interest in motorsports, quite naturally, rubbed off on a certain percentage of the buying public. Soon Ford Model Ts were being stripped down for organized competition at tracks all across the country. Fenderless Ford-backed race cars soon began appearing at the Brickyard in Indianapolis. Ford's association with speed was heightened by the exploits of the flathead V-8 during the 1930s, and soon both racers and bank robbers (such as Clyde Barrow) were making fast getaways in Ford-powered cars.

Postwar Fords continued to find favor with the buying (and hot rodding) public due to their enviable combination of low price and high horsepower. As a result, Ford race cars played a large role in the early rough-and-tumble days of modified "jalopy" racing that directly preceded the advent of the NASCAR tour. And when Big Bill France's genius led to the very first "Strictly Stock" NASCAR race, cars of the Fomoco persuasion constituted nearly half of the starting field (in addition to Roper's car, there were 2 other Lincolns, 10 Fords, and 1 Mercury on the grid when the green flag fell).

As Roper's pre-race commute from Kansas suggests, NASCAR racing was a whole lot different in

Hard charging Curtis Turner won the 1956 Southern 500 in a car just like this. "Pops" qualified his "Purple Hog" 11th that day before going on to best the field at the "Lady in Black."

Fomoco stock cars circa 1956 were motorvated by a beefed-up version of the factory Y-block overhead valve engine. Though not as stout as, say, a Chrysler FirePower Hemi, a full boogie Y-block could still be counted on for in excess of 230 ponies.

the days when Harry Truman was still giving them hell in Foggy Bottom. "Stock" meant stock in 1949, and Roper's car sported just about every piece of sheet metal and brightwork that it had when it rolled off of the assembly line. The same could be said for the 336-cubic-inch straight eight under the car's hood. Chances are that not many more than the engine's original complement of 152 ponies were on hand to propel Roper to victory that day in 1949.

That having been said, it was yet another sleepy straight eight that provided the oomph for Fomoco's second NASCAR win just four months later. That particular race took place on a half-mile dirt oval in Hamburg, New York, in September. Jack White came home in first place after 200 dusty

Stock ergonomics characterized the control cabin of mid-1950's Ford race cars. Stock bench seats and column shifters were part of the program, as were stock dash panels and roll-up windows. Note the wispy roll cage that was all the rules book required in 1956.

circuits of the track in his Moyer Company-sponsored 1949 Lincoln.

When that eight-race inaugural season was over, Oldsmobile driver Red Byron had cinched the very first NASCAR Driver's Championship. He and his fellow Olds drivers won five of the races contested, with Lee Petty's Plymouth accounting for the lone event not won by a GM or Lincoln car. Things would change dramatically in 1950, and the first wins scored by Ford- and Mercury-badged stock cars were just around the corner in the first full decade of what soon came to be called the Grand National series.

Though the first fledgling NASCAR season had consisted of but a handful of short track events, it had attracted more than a little attention from racing fans. Things were definitely looking up for Bill France's new series when cars began filing into Daytona for the 1950 season opener on the famed beach course there. Though steep banking and asphalt pavement were years in the future at Daytona, the 4.17-mile beach and road course just south of town was still a pretty impressive track. It consisted of equal parts surface street and hard-packed sand straights that were connected by treacherously rutted sand and dirt corners. Forty-one cars qualified for the 200-mile, 48-lap event in Daytona that year. In that number were nine Lincolns and seven Fords. Former tank corps driver Harold Kite qualified his #21 Lincoln third on the grid. With a "capacity" crowd of 9,500 spectators looking on, Kite muscled and manhandled his pregnant-looking 4,600-pound "race car" into the lead on lap 1 and went on to win the race by a 53-second margin over Red Byron's Olds. Kite averaged a not-so-blistering 89.894 miles per hour on his way to winning the $1,500 race purse that day.

Tim Flock, who was soon to become a legend on the tour, backed up Kite's Daytona win with a second Lincoln triumph at race number two of 1950 in Charlotte. Though no one knew it at the time, Flock's victory in the Edmund's Motors 1949 Lincoln would prove to be the last Grand National win for the marque in history. But other Fomoco car lines were waiting in the wings to continue the Blue Oval winning tradition.

In fact, Ford win number three took place that same season at Dayton, Ohio. The race in question was a 100-mile event contested on the dirt-covered Dayton Speedway. The green flag fell on 25 cars that day in the Buckeye State, and in that number was the #27 1950 Ford driven by Jimmy Florian for Euclid Motors. Like its assembly line mate the Lincoln, Florian's Ford was powered by a 239.4-cubic-inch flathead V-8 that churned out just 100 horses. Even so, Florian (who in a novel twist raced bare-chested that day to beat the heat!) was able to put each and every one of those ponies to good use during the race. Florian's Ford ran near the front all day and led on two occasions for a total of 40 laps. He took the lead for the final and most important time on lap 168 (of 200 total), when he slipped past the always-hard-to-pass Curtis "Pops" Turner. The Cleveland mechanic kept his Ford out in front for the balance of the race and got to kiss the pretty girl at race's end. He also got to pocket the $1,000 first-place purse.

Florian's win turned out to be the start of the most successful string of wins yet to be recorded by any manufacturer on the NASCAR tour. Over the years that followed, Ford drivers continued to taste victory on the Grand National/Winston Cup circuit. To date, cars carrying the Blue Oval have visited a NASCAR victory lane no fewer than 450 times—more than any other marque. That winning total continues to grow each and every racing season.

Nineteen fifty was also the first year that a car from Ford's sister division, Mercury, visited the winner's circle. That inaugural triumph took place at a 100-mile event held on dirt in Vernon, New York. North Carolinian Bill Blair scored win number one for the Mercury car division on June 18th before a crowd of 15,000 spectators. Lloyd Moore notched a second Mercury victory at the

Ford upped the horsepower ante for 1957 by homologating a special supercharged version of the 312-cubic-inch Y-block engine. Ford engineers rated the engine at 300 ponies in street trim that year. It's a safe bet that ace tuners like Ralph Moody coaxed even higher figures from the littler huffer on race day.

next-to-last race of the 1950 season in Winchester, Indiana. Two more Mercury Man division wins were on tap for 1952.

Racing Dearth for Ford: 1951–1954

Though all three of Fomoco's car divisions had enjoyed some success during the first two seasons of Grand National stock car racing, as things turned out it would be nearly five long years before the next Blue Oval-powered stock car would visit a NASCAR victory lane. Unfortunately, when the factory horsepower wars broke out during the early

1950s, Ford's sleepy little flathead V-8 was found to be lacking in "ammunition." Corporate disinterest in organized motorsports also played a role in keeping Ford cars out of the win column. Other more progressive carmakers like Hudson began to build special factory high-performance components specifically for stock car competition with predictable results—fabulous Hudson Hornets began to win with regularity. Olds drivers found their own advantage with that car line's all-new high-horsepower, overhead-valve V-8s, and they too scored more than a few Grand National victories. Chrysler

Motor Company (Chryco) drivers picked up Hemi power in 1951, and soon they, too, started to show up in the winner's circle. All Fomoco racers could do was stew while General Motors, Mopar, and Hudson were winning in the early 1950s.

The mid-1950s saw the NASCAR tour mature from a marginally organized series consisting of just a handful of races to a mature tour made up of 33 different events. Purses grew with each passing season and so, too, did attendance at tracks along the circuit. Master promoter Bill France staged a Grand National race in Dearborn that was heavily attended by Big Three execs in 1951, and the Southern 500 quickly became a national headline generator.

The increasing public interest (as indicated by burgeoning gate receipts) was not lost on car makers in Detroit (or Dearborn). When Chevrolet's public relations bandwagon began to play up the wins scored by Chevy stock car drivers in 1954, Ford execs simply could not continue to ignore the importance of winning races on Sunday. The General's Chevrolet division had long been Ford's arch rival for the mid-priced car-buying public. While wins scored by high-line Chrysler letter cars and econo-Hudsons could be ignored, when Chevrolet began to rack up high publicity wins in NASCAR circles, the powers that be at Ford just couldn't look the other way. And so it was that Ford decided to venture into factory-backed motorsports in 1955.

It was an event that took place in 1953, however, that brought Ford into victory lane. That was the introduction of the all-new, overhead-valve (OHV) Y-block engine family. The Ford Motor Company began its 51st year of production in 1954 by

While Holman & Moody were working their mechanical magic on the East Coast, Bill Stroppe was turning out trick race cars on the opposite coast. Tim Flock, whose ride is pictured here, drove both convertible and hardtop Mercurys for Stroppe in the 1950s. The cars looked sharp and were plenty fast. *Mike Slade*

17

"M" Power Mercurys came factory equipped with 335 horsepower versions of the corporate Y-block. Two four-barrel carbs, alloy valve covers, and cast-iron exhaust manifolds were all regular parts of the under-hood scenery.

pulling the wraps off of an all-new 130-horse OHV engine. Though displacing nearly the same 239 cubic inches of its predecessor "flatty" V-8, the all-new engine possessed loads of potential and soon grew to 272, then 292, and finally 312 cubic inches (by 1956). The new engine line would soon turn out to be just the solution to the Ford horsepower deficit and to the Blue Oval slump on the track.

Ford Gets Serious About Racing: 1955–1957

For the 1955 season, Ford tapped open-wheel ace Buddy Schuman to supervise their effort, and series hotshoes Curtis Turner and Little Joe Weatherly got the nod for driving chores.

The 1955 Southern 500 was the targeted debut of the Ford-backed, two-car team, and work

got under way for that race with the construction of two Fairlane-based race cars at Ford's experimental garage in Dearborn. In accordance with the rules book of the day, extra shocks were added at all four corners and reinforced rims were installed. A rudimentary single-hoop roll bar (yes, bar, not cage) was installed at about the same time that most of the cars' interior accouterments were removed. Power was provided by a highly-tuned 292 Y-block that was rated at 230 horsepower. A three-on-the-tree manual tranny was all that Ford had in its torque reduction arsenal for 1955, so that's what backed up the willing little small block. When construction was complete, the cars were dressed in a coat of purple Schwam Motors (a Charlotte Ford dealer) racing livery.

During qualifying for the race, Weatherly rocketed around the "Lady in Black" at an average speed of 109.006 miles per hour with Turner a couple of ticks behind at 106. Though not fast enough to capture the pole (which future Ford driver Glenn "Fireball" Roberts snared with a 110-mile-per-hour lap in his Buick), both Weatherly and Turner started towards the front of the pack, and both ultimately spent more than a little time in the lead. Unfortunately, a shunt sidelined Little Joe at mid-race, and a tie rod end put an equally unceremonious end to Turner's day. Herb Thomas' win in Smokey Yunick's 1955 factory-backed Chevrolet must have stung a bit, but Ford wins were just around the corner, specifically around the corner of the Memphis-Arkansas dirt track in LeHi, Arkansas just one month later.

Interestingly, Ford's second win on the Grand National tour was served up not by the Schuman team but rather by Chrysler 300 team owner Karl Kiekhaefer! In addition to sponsoring his usual fleet of all-conquering Chryslers at the LeHi event, Kiekhaefer also backed Speedy Thompson's 1955 Fairlane, and at race's end it was Speedy who took the trophy home. Marvin Panch backed that Ford performance up with a second-place finish in his own #98 Ford.

Like their Ford counterparts, Stroppe-prepped Mercs were basically stock in the control cabin department. And that included a factory-based bench seat (with passenger seat back removed), a stock dash, and factory-installed door panels. Note the air corp gun camera that was one of the first attempts at capturing in-car footage during a race.

Curtis Turner was one of the earliest stars on the NASCAR tour. He earned that fame with the help of a fleet of fast Ford race cars. Here he is at the helm of a supercharged 1957 Fairlane on the beach at Daytona. *Daytona Racing Archives*

Late in the 1955 season, Indy ace Pete DePaolo was lured to the NASCAR ranks to head up Ford's factory effort. He set up shop in Long Beach, California (of all places), and scored one late-season win with the help of driver Buck Baker at Martinsville in October of 1955. When the season finished, Kiekhaefer's Chrysler teams had won just about all of the NASCAR marbles, but Ford drivers had snared two wins and garnered more than a little ink with strong performances in several other events. Things would be even better in 1956.

Preparation for Ford's second official factory-backed season on the Grand National tour got under way with the construction of four Y-block powered

Fairlanes in the Ford "EX" garage at Dearborn. Assembly-line chassis formed the basis for NASCAR stockers in 1956, so the first order of business was removing the various and sundry street-related components that served no purpose in oval-track racing. Chassis reinforcement came next. That included rewelding all the original UAW seams, adding extra metal in high-flex areas, and fabricating extra reinforced shock mounts for the quartet of jounce controllers that were destined for use fore and aft. The final step taken to improve the cars' cornering resolve was the installation of a single loop roll cage.

The 1956 rules book permitted the use of any stock-style brake lining, which, of course, meant that standard drum brakes were fitted to each corner of the chassis. Reinforced rims carrying slender-period bias-ply rubber made the whole package a "roller." Body modifications were almost nonexistent by today's totally fabricated standards. Running lights were removed and their openings were covered; the driver's door was bolted shut and some chrome trim was removed. Beyond that, the team's Ford Fairlanes looked pretty much the same as they had when rolling down the assembly line. Stock was pretty much the theme in the four cars' control cabins, too. Rear seat removal was permitted by the exceedingly thin 1956 rules book, as was 86ing the passenger's seat back on the stock bench seat. Aircraft surplus seat belts were also part of the program, but beyond their addition and that of the rudimentary roll cage, the rest of the cars' ergonomics were unchanged from showroom condition.

Ford stock cars circa 1956 were powered by a newly enlarged and enlivened 312-cubic-inch version of the corporate OHV V-8. In race tune, the 3.80x3.44 bore and stroke engines could be relied on for something in the vicinity of 230 horsepower—barely enough to keep one of the cars up on Daytona's banking today.

When Robert McNamara pulled Ford's racing irons out of the fire in 1957, the newly formed Holman & Moody team elected to soldier on without factory backing. Team efforts were first centered around Y-block-powered Fairlane 500s like this 1958 that Curtis "Pops" Turner campaigned at Charlotte. Note the stock brightwork that Turner carried into battle in the late 1950s. *Daytona Racing Archives*

But by the 1950s' standards, a race-prepped 312's power output made for formidable competition, especially when presided over by Joe Weatherly and Curtis Turner, who reprised their roles as Ford team drivers for the new season. Rising star Fireball Roberts also signed on to drive for DePaolo in 1956, as did New England ace Ralph Moody. All four drivers would become legends in their own right by the end of Ford's Total Performance era (an era that took its name from Ford's performance advertising program). In many ways, that unparalleled period of Blue Oval domination began with their pairing in the 1956 season. DePaolo also assembled a first-rate support staff to keep his fleet of four Fairlanes properly prepared for each race entered by the team. In that number was a husky fellow named John Holman, who had served variously as a mechanic for Bill Stroppe's West Coast-based Mercury team and as team truck driver for Lincoln's legendary Carrera Pan Americana Mexican Road Race team. Ford fans will, of course, immediately understand how significant that first pairing of John Holman and Ralph Moody (not to mention Turner, Weatherly, and Roberts) ultimately turned out to be since the now fabled Holman & Moody car-building team was a direct result.

Ford's first victory of the official 1956 season had come in the waning months of 1955, when AAA champ Chuck Stevenson clinched the road course event on the as-yet-unpaved Willow Springs Speedway in California (the Triple A served as a competing sanctioning body to France's

Holman & Moody began to build customer cars in 1958 and 1959 out of bare "Square Bird" chassis that had been "discarded." The wins scored by those Thunderbirds were the first in a long line of victories that continue today. Pictured here at speed in Daytona in 1959 are drivers Tom Pistone, Bob Welborn, and Curtis Turner. *JDC Collection*

NASCAR organization in the 1950s). That win was a portent of the success waiting for Blue Oval and Mercury drivers in the season ahead.

The first race of the 1956 calendar year—the beach race in Daytona—took place in February, and during the straight-line speed runs that constituted qualifying for that event, the new Ford contingent made its presence known. Moody translated a 22-second starting berth into a more than competitive third place finish during the 160-mile beach race at the not yet "Big" D—just behind Billy Meyers' second in a Stroppe-prepped Mercury. On his way to the flag, Moody completely flipped his #12 Ford but pressed on undaunted when he landed on his wheels in the soft sand!

One month later, Meyers parked his 312 motorvated Merc in the winners' circle at West Palm Beach to score the first Fomoco win of the calendar year. Moody scored his first victory for the DePaolo team in June at LeHi, Arkansas, and that victory ended the incredible string of 16 straight wins for Kiekhaefer's Chrysler teams. Fireball Roberts notched the next team win in a 250-miler at Raleigh in July and then again in a 100-mile event at Chicago two weeks later. Curtis Turner scored arguably the team's most significant win of the season at Darlington in September, when he outpaced a 70-car field to claim the Southern 500. Moody and Roberts conspired to produce a total of 9 team wins in addition to Turner's triumph at the "Lady in Black." Weatherly's 6 top-five finishes helped add to a team total of 40 in that category—not too shabby for the team's first full season of competitive effort.

All told, when factoring in the victories scored by Bill Stroppe's West Coast Mercury team and the wins recorded by independently-backed Ford and Mercury drivers, Fomoco's 1956 tally was 19 wins in 56 events. Kiekhaefer's 300Cs, which dominated the season with a total of 22 trips to the winner's circle, were the only cars to visit that hallowed spot more frequently than the Ford and Mercury contingent.

All in all, things looked quite rosy for Fomoco's prospects in 1957.

Ford racers were also enjoying more than a little success in NASCAR's sister "convertible" division during the 1956 season. Turner (serving double duty in both the convertible and Grand National divisions) won the beach race that served as the "undercard" event at Daytona in 1956. Weatherly and Roberts also double-dipped in the convertible division that season, and all three of them found that some of their stiffest "zipper top" (as convertible stockers were referred to because they were Grand National cars with their tops unbolted) competition was meted out by a young fellow from Virginia named Glenn Wood, whose #22 Ford was prepared by his younger brother Leonard. The 1956 convertible series produced a total of 26 Ford wins (of 45 events contested) and provided the impetus for the now legendary Wood Brothers Grand National team that still competes on the NASCAR tour today.

It's a pretty safe bet that folks in the Dearborn "Glass House" were pleased with the results of their first full season of NASCAR stock car competition. Money and midnight oil soon began to be expended in pursuit of even more stock car success in the coming season. Soon, special power-train pieces ostensibly designed for export and police work began to pop out of Blue Oval casting boxes. Their intended destination was, of course, the NASCAR circuit. And so, too, was the special Thunderbird and Fairlane McCulloch supercharger engine package that Ford rolled out for the 1957 model year. Rated at 300 horsepower (340 in race trim), the new force-fed induction system was the last word in high tech for 1957 and more than trumped Chevrolet's 1-horsepower-per-cubic-inch, fuel-injected 283 small block.

On the personnel side, Ford extended offers of employment to ace mechanic Smokey Yunick and up-and-coming driver Marvin Panch. Yunick's previous relationship with hotshot driver Paul Goldsmith

Nineteen fifty-eight was the year that Ford rolled out its all new big-block FE engine family. Displacements that year were modest but still good enough to produce 350 horsepower. By 1963, Ford engineers had perfected the 427 FE engine that went on to dominate the 1960s' scene both on and off the track.

ensured that he, too, would be making an appearance behind the wheel of a Fomoco car in 1957.

As was the custom in those days, the 1957 Grand National season got under way in late 1956, with race one running at Willow Springs in California. New DePaolo driver Marvin Panch left the field in his dust (literally, as the road course there was still dirt) and won the race from the pole position. Fireball Roberts came home second that day to further darken Chevrolet's prospects for the coming season. Panch, Roberts, Moody, and Goldsmith then went on a tear through the Grand National ranks, winning 13 of the next 19 events. Making matters even worse for Chevrolet and Mopar drivers were

the 3 additional Ford wins scored by independent drivers along the way.

John Holman's promotion to the directorship of DePaolo's team just into the season undoubtedly had a positive effect on Ford's winning streak. His natural leadership ability gave a focus to Ford's efforts that some had felt was lacking when DePaolo alone had been charting the course.

AMA Ban/Holman & Moody Set Up Shop: 1957–1960

Unfortunately, trouble was just around the corner for the Ford juggernaut. That trouble came in the form of a ban on factory-backed motorsports competition that was enacted by the Automobile Manufacturer's Association (AMA) in June of 1957. In reality, the ban had been engineered behind the scenes by GM exec Harlow "Red" Curtice. Realizing that Ford's continued high-profile rout of the GM teams was bound to have a deleterious effect on The General's market share (and perhaps aware of just how gullible Ford General Manager Robert McNamara really was), Curtice got the AMA to ban all factory-backed racing efforts. When, as expected, McNamara took the edict seriously and shut down all Fomoco racing efforts, Chevrolet teams (who were still clandestinely receiving factory support) were able to reverse the tide of Blue Oval wins. Whereas before the ban Ford drivers had won 15 of 21 events contested, they were only able to capture 12 of 32 after it went into effect.

As history records, the faucet of Ford funding dollars remained turned off until McNamara left to become the Kennedy Administration's secretary of defense, searching for light at the end of the Vietnam tunnel. If any good did come out of the AMA ban for Ford fans, it was the resulting pairing of Ralph Moody and John Holman in Charlotte, North Carolina. When McNamara closed down DePaolo's operation, then-racing-chief Jacque Passino pulled the strings that made it possible for Holman & Moody to set up shop with the leftover parts already in the funding pipeline. Joe Weatherly fielded the very first H&M-sponsored car in July 1957 at the Raleigh 250 in Raleigh, North Carolina, where he started 10th and finished 3rd.

With Ford officially on the sidelines, the 1950s ended with a fizzle for Ford fans of Grand National stock car racing. But Holman & Moody soldiered on and, in a visionary move, began to build "customer" turn-key race cars. In an era when drivers normally built their own competition mounts (save for the lucky few with factory backing), it was indeed novel for Holman & Moody to begin building competition cars for the bucks-up buying public. Some of those first customer cars were built out

of "scrapped" Thunderbird unit body chassis that Holman had been able to back-door out of Dearborn for a pittance. When fitted with a race spec version of the all-new FE (Ford Engine) big-block engine introduced in 1958, the lightweight little "Squarebirds" were indeed formidable mounts—factory-backed or not.

The NASCAR record book shows that Holman & Moody team win number one came at Champion Speedway in Fayetteville, North Carolina, in March 1958. Curtis Turner had stayed on as an H&M driver even after McNamara had packed Ford's racing tent, and his unparalleled skill as a short-track driver no doubt played a role in the 50-mile feature victory. It was to be but the first of many Ford and Mercury wins scored by a Holman & Moody-prepped racing car. Turner added 2 more wins to the H&M victory column later that season. Other Ford wins came courtesy of Junior Johnson (6, in fact), Parnelli Jones, Joe Weatherly, Jim Reed, and Shorty Rollins, but Chevrolet drivers carried the day (and the season) with 23 Grand National victories.

Unfortunately, things were destined to get worse before they would get better for Fomoco drivers and fans along the NASCAR circuit. Nineteen fifty-nine produced a new low in the number of Blue Oval visits to victory lane, and, all told, just 10 Ford (and no Mercury) wins were recorded that year.

Though an off-year for Ford teams, things were afoot behind the scenes (and on the surface streets of America) that would soon return factory drivers to the position of total domination they'd enjoyed just before the 1957 AMA ban went into effect. And those forces were driven in large part by the buying public. Though Ford had retreated from racing, it had not lost track of what the automotive buying public wanted in the way of performance—and that was horsepower. Ford answered that demand by quickly increasing the size of its corporate big block from 332 to 352 cubic inches. Each pass with the boring bar produced even more horsepower, and quite naturally, independent drivers quickly put those ponies to work in the NASCAR ranks.

Aiding in their quest for victory was the slippery, fast-backed shape of the all-new Galaxie Starliner line that was introduced in 1960. In race trim, a 360-horse 1960 Starliner was good for speeds in excess of 145 miles per hour on big tracks like Daytona. The car's swoopy shape helped out at lesser tracks, too, and as a result, Ford drivers scored 15 wins that season. Joe Weatherly is credited with 3 of those triumphs for the Holman & Moody team, and the Wood Brothers scored their first ever Grand National wins (including the National 500 at Charlotte) that season. Another

future Fomoco champion, Ned Jarrett, also recorded Ford wins in 1960.

Nonetheless, with Ford still officially out of racing and its GM rivals still very much in it, Chevrolet and Pontiac drivers pretty much ran away with the 44-race series (20 total wins, plus 1 for Oldsmobile).

At season's end, Lee Iacocca was elevated to the general manager post at Ford (in place of the not-so-dearly-departed McNamara), and Ralph Moody signed a promising young Midwestern driver named Fred Lorenzen to an H&M contract for the upcoming season. Better yet, Ford announced that it was breaking out the boring bar once again, and for 1961, the FE block would displace a full 390 cubic inches. Things were looking up!

A boxy Thunderbird just like this one took the checkered flag first (or so it seemed) at the inaugural Daytona 500 in 1959. T-Bird driver Johnny Beauchamp was initially declared the winner of that event and got to kiss the pretty girl in victory lane. Unfortunately, photo analysis of the finish four days later ultimately revealed that Lee Petty's Oldsmobile actually had inched Beauchamp's car out at the line. Curtis Turner scored the first of many Thunderbird wins one month later at Hillsboro, North Carolina.

Nineteen sixty-nine was the first season of the aero wars. Fomoco combatants rolled into battle at the helm of stretch-nosed Mercury Cyclone Spoiler IIs and Torino Talladegas. By season's end, drivers like LeeRoy Yarbrough (#98) and Richard Petty (#43) had pummeled their winged and wingless rivals into submission with their superior aerodynamics. *Courtesy Ford Motor Company*

2

Ford Returns with Total Performance

Iacocca Takes the Helm: 1961

Lee Iacocca's ascension to "Glass House" heights was a breath of fresh air for performance-oriented Ford engineers. Rising as he did from the sales ranks, Iacocca was well aware of the effect that Sunday wins had on Monday sales floor traffic. And so it was that the doors that McNamara had closed to Ford racers in 1957 began to open up just a bit.

Fast Freddie Lorenzen scored his first Holman & Moody win at Martinsville in the 1961 Virginia 500. His trademark white and blue #28 Galaxie had started second on the grid before going on to lead the event from lap 118 to its conclusion on lap 149. Lorenzen's next win came at Darlington in the Rebel 300, where his 375-horse, 390-powered Starliner started on the pole. Lorenzen also won at Atlanta for H&M.

However, the biggest Holman (and Fomoco) win scored in 1961 was, without doubt, the triumph recorded by Nelson Stacy in the headline-producing Southern 500. The pugnacious Cincinnati native scored the upset victory by passing Plymouth driver Marvin Panch with just seven laps to go before a crowd of 80,000 cheering fans.

Ford Resumes Factory Backing: 1962

Nineteen sixty-two was the season that Ford dispensed with the AMA ban altogether and announced its return to fully factory-backed sponsorship of motorsports competition. Interestingly, that move was sparked in part by poor aerodynamics. Though 1960 and 1961 Galaxies could be bought with smooth-flowing Starliner roof panels, for 1962 the squared-off look was in,

1961–1970

and Galaxies sported an upright, drag-producing back light. It was a step backwards in aerodynamics that the extra cubic inches carried by the new 406-cubic-inch engine couldn't overcome. When Fireball Roberts swept Speedweeks in Daytona that year with his convertible rooflined (and 421 Super Duty-powered) Catalina, Ford racers looked to Dearborn for help.

Their requests were answered just before the Atlanta race with a new factory Ford "option" called the Starlift roof. Basically a removable version of the previous Starliner top panel, the new option supposedly was intended to bolt onto a Galaxie convertible in street application. Problem was, when that was attempted, the windows wouldn't close all the way (you see, there was no Starlift window option). It was obvious to all that the new lid was intended specifically for the NASCAR circuit. And that placed Bill France and his sanctioning body in a bit of a bind. After all, they'd

Why is this man smiling? It could be because he is Ralph Moody, and he knows just how dominant the race cars he builds will be throughout the 1960s. The Starliner he's leaning on is an early example of Ford aerodynamics. Like the Talladegas, Spoiler IIs, and Thunderbirds that would follow it onto the track, a 1960/1961 Starliner owed much of its speed to its slippery shape. *Daytona Racing Archives*

Nelson Stacy was an early Holman & Moody team driver. He piloted his #29 H&M-prepped race cars with the same tenacity he'd displayed as a tank driver in World War II. *Daytona Racing Archives*

been quietly looking the other way when Chevrolet and Pontiac offered up over-the-counter "stock" options (like the crate Super Duty 421s that Poncho drivers had been whipping the competition with), so how could they nay-say Ford's new roof?

So at first they didn't. As a result, Lorenzen and Stacy turned up at the Atlanta 500 with Holman & Moody-prepped Starlift Galaxies and stole the show. Fast Freddie took what turned out to be the body style's only NASCAR win. In a move dripping with more than a little hypocrisy, NASCAR banned the new roof panel just after the race. That act prompted Henry Ford II (the Deuce) to announce shortly thereafter that the Ford Motor Company would no longer be honoring the AMA

ban on factory-backed motorsports competition. Holman & Moody was put back on the company payroll in short order, and big plans were laid for the future. Not the least among those was a proposed version of the tried and true FE engine that had been punched out to an estimable 427 cubic inches. Though 1962 produced just six Grand National wins (in that number a Southern 500 triumph by Larry Frank), things would be quite different the very next season.

A Phenomenal Season: 1963

During the off season, Ford stylists cooked up an all-new Galaxie body that was much better at cutting through the wind than its predecessors had been.

31

Though different in just about every regard than the boxy 1962s it replaced, the new 1963 was most changed in the roofline area. Instead of the upright back light of the year before, the new full-sized cars featured a sporty looking "convertible" roofline that was the next best thing to the Starlift panels of yore. Power for the slippery new chassis was provided by a new and improved version of the FE that boasted 427 cubic inches and 410 horsepower. Blessed with a race-proved, cross-bolted bottom end and free-flowing "Low Riser" heads, the new oversquare big block was both reliable and powerful at high rpm. Dan Gurney proved just that at Riverside in a Holman & Moody-prepped Ford when he won the Riverside 500 in January of 1963.

Daytona was next on the schedule, and Ford drivers had high hopes as they pulled into the garage area at the Big D. Until, that is, Junior Johnson and his Bow Tie stablemates fired up their all-new and decidedly non-production 427 Mystery Motor engines for the very first time. Though Big Bill France had made a show of banning the 1962 Starlift option for being a non-production item, he suddenly decided to look the other way when Chevrolet drivers showed up with an all-new non-production polyangle-valved 427 big block (of which ultimately only 48 or so were ever cast up).

Fomoco forces were very unhappy with the situation until reliability problems sidelined the new porcupine-powered Chevrolets. Ford fans' anger at the sanctioning body's duplicity was more than a little assuaged when Iowa native Tiny Lund scored the very first Daytona 500 win for the Blue Oval in

Glenn "Fireball" Roberts was one of the first Holman & Moody team drivers in 1956. Unfortunately, Ford's competition pull-out forced him to switch to GM cars in 1957. Fireball finally returned to the Ford fold in 1963 to campaign this Passino Purple "Pontiac Eater" built by Holman & Moody. He proved he still knew how to pilot a fast Ford by winning his very first Galaxie outing that season. *Mike Slade*

West Coast racer Bill Stroppe built some of the best-looking race cars to ever grace a Grand National racing grid. Stroppe was fond of red, white, and blue and painted his team Mercurys in those hues in the 1950s and 1960s. Darel Dieringer drove this #16 Marauder for Stroppe in 1964.

storybook fashion. Lund had saved Marvin Panch's life by yanking him bodily from a burning sports car wreck just days before the 500 (Tiny was a small man in name alone). Panch had rewarded that brave act by nominating Tiny for the Wood Brothers Galaxie ride he'd have to give up while recuperating. As mentioned, Tiny crossed the finish line first, followed by Fast Freddie, Ned Jarrett, Nelson Stacy, and Dan Gurney who were also all mounted in 427 Low Riser-powered Galaxies.

Lorenzen's performance during the 1963 season was particularly memorable. He backed up his second-place finish at Daytona with wins in the Atlanta 500, the World 600 (at Charlotte), the Volunteer 500 (at Bristol), the Western Carolina 500 (at Weaverville), the Mountaineer 500 (at Huntington),

and the Old Dominion 500 (at Martinsville). Lorenzen's success earned him $122,587.28 in winnings. It was the first time that any Grand National driver had broken the six-figure mark.

Ford fans will also fondly remember that 1963 was the year that Fireball Roberts returned to the Ford fold in a "Passino Purple" #22 H&M Galaxie. Fireball won his first 1963 Ford event at Bristol in the Southeastern 500. He went on to score three more Ford wins that season, the most important of which was a victory in the 1963 Southern 500, which ran unimpeded without a single caution. It was a convincing win to say the least, as Fireball had set the fastest lap in qualifying (at 133.648) before going on to win at an average speed of 129.784 miles per hour. Making the win even sweeter for Fomoco fans was the fact that Marvin Panch (Ford), Fred Lorenzen (Ford), Nelson Stacy (Ford), Darel Dieringer (Mercury), Rex White (Mercury), Joe Weatherly (Mercury), and Tiny Lund (Ford) had freight trained across the Darlington finish line in Fireball's wake. Total Performance indeed!

Ford drivers won 23 events in 1963. And that was more than any other manufacturer. Nineteen sixty-three was also the year that Mercury returned to the win column after a six-year hiatus. Darel Dieringer served up those honors at the last race of the season, the Golden State 400 at Riverside, where he piloted his Bill Stroppe-prepped red, white, and blue Marauder to victory lane. That particular Riverside event also marked the arrival of Bud Moore to the Fomoco fold. It was the beginning of the association between Moore's Spartanburg-based team and cars of the Fomoco persuasion that continues today.

Ford and Chrysler Battle It Out (Hemi Supremacy): 1964–1965

There's no doubt about it, 1963 was a phenomenal season for Ford's Total Performance team. As a result, it's pretty likely that Ford and Mercury teams were looking forward to more of the same for 1964. Especially so, since GM had pulled up stakes and withdrawn from racing during the 1963 season for fear that its continued clandestine (in the face of the AMA ban) racing activities would result in antitrust litigation from the feds.

But things don't always turn out as planned. Fomoco drivers began to understand that fact (in spades) the very first time that Richard Petty fired up his all-new 426 Chrysler Hemi engine in the Daytona garage area during Speedweeks 1964. Like the year before, Ford forces found themselves confronted by an all-new, not-in-the-least-production engine that the sanctioning body had deemed legal for Grand National competition.

Nineteen sixty-three Marauders shared the same convertible-style roofline that their Galaxie counterparts carried. It was as much a concession to styling concerns as to speed. One thing is for sure, the swoopy new roof worked a whole lot better than the vertical-back-light, formal roof panels of 1962 ever had. *Courtesy Bill Holder*

Though Ford's top brass howled about the fact that no regular production Hemi cars were slated for production (and indeed wouldn't be for two full years) NASCAR officials turned a deaf ear and gave the purpose-built racing engine their stamp of approval. And unlike the previous year, the formidable new foe would not be plagued by the same type of teething problems that had hamstrung the Chevrolet Mystery Motor.

Proof of that fact became all too clear at Daytona during qualifying where Petty and his Mopar minions ran away with both qualifying and the race. When the checkered flag fell, it was Petty first, followed by two other Plymouths. Adding to

Fred Lorenzen was a midwesterner who caught Ralph Moody's eye during the 1960 NASCAR season. He signed on as an H&M team driver the next year, and the rest is history. During the 1960s, Lorenzen became NASCAR's Golden Boy. He won 26 Grand National events and was the first driver to win more than $100,000 in a single season. Fast Freddie is posing inside the Holman & Moody shop with his new "fastback" 1963 Galaxie stocker. That's legendary mechanic Herb Nab leaning on the car's windscreen. *Daytona Racing Archives*

the funk that Ford drivers found themselves in was the pall that had fallen over the series ever since Joe Weatherly had been killed in the Motor Trend 500 while at the wheel of a Bud Moore Mercury. Even NASCAR's decision to allow all Fomoco cars to run special High Riser head castings atop their 427 engines had not been enough to brighten the Ford Teams' outlook—especially after Daytona.

Things got even worse at the World 600 in May when Fireball Roberts became involved in a shunt with fellow Ford drivers Ned Jarrett and Junior Johnson. Tragically, the still evolving NASCAR rules book did not require fuel cells (nor truth be known, did any working examples yet exist at the time). As a result, when Fireball's purple Galaxie got turned upside down and burst into flames on the backstretch, Fireball was doomed. Horribly burned before he could be pulled from the wreck, he lingered until just days before the Firecracker 400 in July.

Though Ford and Mercury drivers notched a total of 35 victories in 1964 (including triumphs at Atlanta, Wilkesboro, Martinsville, Darlington, Charlotte, Bristol, and Watkins Glen), the 1964 season was not generally a happy one for Ford teams. Petty's 1964 driving championship suggested that something would have to be done—and fast—if Ford was to have any chance of recapturing the Grand National spotlight.

Chrysler Boycotts: 1965

Ford responded with the 427 single overhead cam (SOHC) engine. When Petty drove home (literally) the supremacy of the hemispherical combustion chamber, Fomoco engine and foundry engineers wasted little time in coming up with a Hemi design of their own. But not just any Hemi, you understand. Their version featured a free-revving head casting that sported one overhead-mounted cam per cylinder and sewer-pipe-sized passages leading in and out of the chambers. Horsepower was predicted to be in the 650 range.

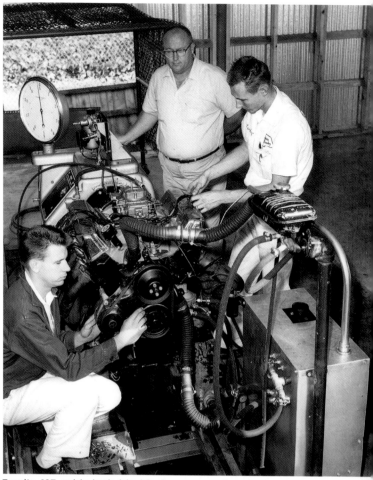

Ford's 427-cubic-inch big block came on line during the 1963 season and saw service well into the 1970s. Though not graced with hemispherical combustion chambers, the high-revving FE was still able to crank out more than 600 horsepower with ease. Here John Holman oversees a 427 dyno run. That's an incredibly young Waddell Wilson working on the front damper. *Daytona Racing Archives*

Following pages
Though not as sleek as the Galaxies they replaced, the restyled Ford full-sized line in 1964 was still more than a little fetching. The race cars that sprang from that production run were fast to boot. Tiny Lund drove an independently-backed 1964 during that period, which has been recently returned to racing trim.

37

The NASCAR rules book in 1964 was far less stringent than the one in use today. Note, for example, the single driver's side roll cage bar that was all the sanctioning body required. Production-based bucket seats like the one installed in this Bill Stroppe Marauder were also part of the program that season.

Stock cars were still pretty stock circa 1964. Note the factory dash panel, the production-based bucket, and the full roll-up windows in Tiny Lund's 1964 Galaxie.

Chrysler got wind of the new SOHC motor and fired back its response in the form of a dual overhead cam version of the 426 that had the potential to produce more than 900 ponies in race trim.

Bill France took in the fracas from the sidelines, and as the 1965 season approached, he decided to act. Shortly before the new year, France announced that safety (and not lack of regular production status) required the outlawing of both the 426 Hemi engine and the concessionary Ford High Riser head castings. Chryco teams promptly announced a boycott of the series that no amount of counternegotiation could undo.

As you might have guessed, 1965 became the year that Fomoco drivers won just about everything there was to win on the Grand National circuit. Fred Lorenzen started that successful season with a rain-shortened triumph at the Daytona 500, and then he and his Blue Oval buddies went on to win an incredible 48 of 55 events contested—including every single superspeedway race on the schedule.

Marvin Panch backed up Lorenzen's win at Daytona by leading a five-car Ford sweep across the stripe in the Atlanta 500 in April. Junior Johnson, just a handful of races away from hanging up his driving gloves, proved he still knew the way to the winner's circle at the Rebel 300 at Darlington in May. Lorenzen took the checkered flag first in the grueling World 600 on Memorial Day. Panch and the Wood Brothers put their name on a deed of ownership at Atlanta by winning the Dixie 500 in June, and the Woods tasted victory again at Daytona when a car they set up for A. J. Foyt bested all comers in the Firecracker 400. Ned Jarrett proved he could win more than just short-track events at Darlington in the Southern 500 where the soon to be two-time NASCAR champ cruised to a whopping 14-lap margin of victory.

Fast Freddie put a Holman & Moody lock on victory lane at Charlotte when he won the National 400 in October, and Curtis Turner rounded out Ford's (and the Wood Brothers') stellar superspeedway season with a triumph at Rockingham in the American 500. All in all, the only way to get a look at the checkered flag in 1965 was from behind the wheel of a Grand National Galaxie. Ford short-track winners that season included Junior Johnson, Tiny Lund, Dan Gurney, Marvin Panch, Dick Hutcherson, Cale Yarborough, A. J. Foyt, and the eventual season champ, Ned Jarrett. Darel Dieringer gave Mercury fans something to cheer about when he won one of the Daytona qualifiers in a Bud Moore-built 1964 Marauder. Though Dodge and Plymouth drivers did trickle back to the series in the waning days of the

season, it was too late for them to undo what has to be the most successful season ever enjoyed by a car manufacturer on the NASCAR tour.

Perhaps as important as the string of victories scored by Fords in 1965 was Ralph Moody's perfection of the Grand National chassis that took place that same year. Though the cars that rolled out of the H&M shop that season had all begun as UAW assembly-line stockers, the chassis mods that mechanical genius Moody outfitted them with turned out to be the industry standard for the next decade and a half—for all cars, regardless of make or manufacturer. Starting with a stock coil-spring

When the sanctioning body permitted Chryco racers to field non-production Hemi engines in 1964, they threw Ford racers a bone in the form of equally non-production High Riser (HR) heads. Though legalizing the HR castings helped a bit, it didn't balance the books entirely. As a result, 1964 was an off year for Ford teams on the tour.

Fred Lorenzen was the class of the 1965 field in his #28 Holman & Moody-prepped Galaxie. Twenty-two years later, Davey Allison was able to make that same boast with the help of the #28 Robert Yates-prepped T-Bird. Both cars now belong to North Carolina's Kim Haynes.

1965 Galaxie chassis, Moody first installed screw-jack-adjustable spring mounts at all four corners. A four-link trailing-arm, live-axle rear suspension based on the Ford 9-inch differential came next, along with a special handmade rear cross-member. A cross-chassis sway bar and a Watts link were then mounted fore and aft respectively, along with enough shock mounts to accommodate a pair of shocks at each wheel. Brakes consisted of massive finned 11x3 drums that were acted on by reinforced, fully-metallic shoes. When fitted with a six-point roll cage, Moody's 1965 Galaxie chassis wasn't all that removed from Cup cars of the modern era. In fact, the rear-steer cars that continue to compete on the circuit still carry what is essentially a 1965 Galaxie chassis (albeit a fully fabricated one) beneath their slippery flanks.

Ford Boycotts: 1966

NASCAR's late season concessions to the Mopar hordes in 1965 had allowed them once again to run the 426 Hemi engine in lightweight

Fast Freddie Lorenzen carried on as a Holman & Moody team driver in 1965. His #28 Galaxie signaled Ford's dominance that season with a win in the Daytona 500. Kim Haynes has recently returned one of Lorenzen's cars to pristine condition.

Dodge and Plymouth unit-body cars. Those same rules were slated to govern in 1966, too, and Chryco was planning to eliminate all doubt about the Hemi's production status by finally offering it for sale as an RPO option. Ford's response was to again seek approval of the 427 SOHC engine. Much to the corporation's disappointment, Big Bill and the sanctioning body rebuffed all of Ford's efforts to that end, even after Ford announced that it, too, was building regular production versions of the 427 Cammer (which turned out to be not true).

When France hung tough, it was Ford's turn to boycott the Grand National series—a step taken after Ford Galaxie teams were trounced by Hemi cars at the Daytona 500. (Richard Petty finished

Hutcherson took in the rapidly passing scenery from this perch in 1965. Note the add-on bolster that more (or less) transformed a regular production bucket into a racing seat.

Hutcherson swept out of the midwestern corn fields in a cloud of dirt-track dust to become a Holman & Moody team driver in the mid-1960s. In 1965, he campaigned this block-long Galaxie on both dirt and asphalt.

All of Fred Lorenzen's H&M cars carried the words "Think! W.H.M." on the passenger side of the dash panel. That script was put there originally by Ralph Moody to keep Fast Freddie focused on "What the Hell Matters." Other "details" of note in a 1967 Fairlane control cabin such as this include the stock dash pad, production-based seat, and RPO steering wheel. Also, if you look closely, you can see the roll cage-mounted strobe light that driver Lorenzen used (in concert with a floor-mounted trap door) to check on right front tire wear during a race. When in use, the strobe "froze" a section of tread for visual inspection.

one lap ahead of Cale Yarborough's Banjo Matthews-prepped Galaxie.) Beginning with the short track race at Hickory in April, Ford teams, for the most part, elected to stay home on Sunday afternoons and watch football. A few independent

When Ford decided to boycott the 1966 season, not every Fomoco team elected to follow suit. Bud Moore, for example, opted to continue to enter selected events. By the middle of the season, he'd begun to field lithe little Comet race cars that had been built out of a combination of 1965 Galaxie (front) and reinforced unit body (rear) suspension components. The end result was the first Fomoco "half chassis" car. Moore's win with driver Darel Dieringer at the Southern 500 helped convince other Ford racers to start racing intermediates upon their return from the boycott late in the season. *JDC Collection*

thinkers like Tiger Tom Pistone, Junior Johnson, and Bud Moore did make appearances at a handful of events—with NASCAR's decided encouragement. It is perhaps those appearances that ultimately had more significance for Ford racers than the losses conceded by the factory-backed boycott.

Junior Johnson, for example, showed up at the Dixie 400 in Atlanta with a radically modified 1966 Galaxie that had been so twisted and distorted away from stock in the pursuit of improved aerodynamics that wags in the garage area took to calling the yellow car the "Banana." NASCAR let the car run (along with an equally novel-bodied Chevelle built by Smokey Yunick), and the car's creative sheet metal led directly to the first use of the now-common body templates. Bud Moore's diminutive little Mercury Comet also made a handful of very

Ford teams elected to run destroked 427 Tunnel Port engines at some tracks during the 1967 and 1968 seasons. Horsepower figures were still high enough to produce 27 Torino and Cyclone wins that year. Cale Yarborough (#21) and David Pearson (#17) were both regular fixtures in victory lane in 1968. *JDC Collection*

significant appearances on the tour during the boycott. Built not from a full-sized, full-frame Mercury but instead from a much smaller, unit-bodied Comet, Moore's #16 car was more than a little innovative. It was, in fact, the wave of Ford's future and the beginning of the end for the Galaxie line in NASCAR circles. Like its Hemi rivals, the red, white, and black car that Moore built for Darel Dieringer featured a unit-body chassis that had been beefed up for use with a 427 Medium Riser engine. When fitted with Galaxie suspension components and the front half of a Galaxie frame, the little Comet became a formidable race car—so much, in fact, that Dieringer bested all comers at the grueling 1966 Southern 500, even without the assistance of a Hemi engine under the hood.

When Fomoco drivers trickled back to the tour late in the season (just as their Chryco rivals had done the year before), they too rode into battle in intermediates that had been outfitted with "half chassis" Galaxie underpinnings and firebreathing Medium Riser big blocks. Fred Lorenzen used just such a car to win the Old Dominion 500 at Martinsville in September upon the Holman & Moody team's return to the circuit. Dick Hutcherson also found his way to victory lane late in the season in a scaled-down Fairlane race car. Even so, those late season triumphs (Ford and Mercury drivers notched a total of 12 Grand National wins in 1966) could not come close to matching the 24 triumphs turned in by Hemi drivers during the boycott.

A Sign of Good Things to Come: 1967

The Cammer was still unwelcome on the NASCAR circuit for 1967, but NASCAR tried to level the playing field by permitting Fomoco racers to install a pair of deep-breathing Tunnel Port head castings atop their 427 short blocks. Tunnel Port heads picked up that moniker due to the fact that their intake passages were so large that the engine's pushrods had to run right through the middle of them (in specially sealed tubes). Horsepower figures climbed significantly with their use, but as the NASCAR record book reflects, it wasn't enough to slow the Hemi-car stronghold on victory lane.

Parnelli Jones got the year off to a good start with a Ford win at Riverside in January, and Indy ace Mario Andretti made a "guest" appearance at the Daytona 500 in February that resulted in a trip to the winner's circle for his #11 Holman & Moody-built TP Fairlane. Fred Lorenzen's win in one of the pre-race "Twin" qualifiers, coupled with his second-place finish in the 500, seemed to portend a successful year for Fomoco teams on the tour. Unfortunately for Blue Oval lovers, 1967 was the season that Richard Petty was crowned the King of Stock Car Racing by winning a phenomenal 27 Grand National races. All told, Petty and his Plymouth and Dodge stablemates won a dominating 36 of 49 events they contested.

The new Mercury intermediates that debuted in 1968 were the last word in aerodynamics. Blessed with a swoopy fastback roofline and a sleek-angled hood line, the cars were fast right off of the trailer. And that's just what similarly named (but unrelated) drivers LeeRoy Yarbrough and Cale Yarborough proved at superspeedways all season. *JDC Collection*

The big news for Ford racers in 1968 was what was behind them—specifically, the swoopy, fastback roof panels. The all-new aerodynamic silhouette on cars like this A. J. Foyt/Jack Bowsher Torino made them nearly impossible to beat that season.

But as bleak as 1967 was overall for Ford and Mercury drivers, there were a few bright spots that foreshadowed the successes of the upcoming season. Darel Dieringer led from the pole at the Gwyn Staley 400 at North Wilkesboro to record Junior Johnson's first Grand National win as a car owner. Interestingly, Johnson attributed the win to his decision to outfit his #26 Fairlane with a destroked, 374-cubic-inch version of the 427 Tunnel Port that provided greater reliability than the full-sized version of the engine. Dick Hutcherson, in his last full season as a driver, also scored wins for Ford (and car owner Bondy Long) in the Smokey Mountain 200 and the Dixie 500. Bobby Allison accounted for a couple of

Fomoco victories, himself, when he used his Holman & Moody #11 car to win the last two races of the year at Rockingham and North Wilkesboro.

Cale Yarborough scored two other Ford wins that year—at the Atlanta 500 and the Firecracker 400—that in many ways served as a preview of the success that he and the Wood Brothers team would enjoy just one season into the future. And that phenomenal year got underway at Daytona in February of 1968.

Aero Wars: 1968–1970

As was the custom in the 1960s, the official beginning of the 1968 Grand National season was in

November of 1967. Bobby Allison won race number one for Holman & Moody in a 1967 Fairlane at Macon, and Dan Gurney continued his dominance of the Riverside Road course in a Wood Brothers-prepped 1968 Ford. But for all practical purposes, the "real" 1968 season began when the Wood Brothers parked their all-new fastback Mercury Cyclone in the garage area at Daytona just prior to the 500.

Nineteen sixty-eight was a model-change year for Ford and Mercury car lines, and the new year brought with it an all-new intermediate body style that looked fast even while standing still. Principally responsible for that illusion of speed was an exaggerated fastback roofline that began at the "A" pillars and swept in an unbroken arc to the cars' rear deck lids.

Mercury referred to its new intermediate variously as the Montego and the Cyclone. Ford intermediates soldiered on behind the Fairlane moniker, and high-line iterations came to be called Torinos. Whatever they were called, the two new intermediates were fast. Significantly faster, in fact, than the stubby little 1967s they had replaced—even when powered by the same twin-four-barrel inducted Tunnel Port engines.

Bud Moore returned to stock car racing in 1968 after a one-year stint in Trans-Am. His driver that season was the popular Tiny Lund. Moore's race car of choice for 1968 continued to be a Mercury. Check out the stock door handles that were still required by the rules book that season. *JDC Collection*

Behind the wheel of a 1968 Mercury Cyclone, Cale Yarborough rocketed around Daytona at 189.222 miles per hour during qualifying for the 500 and won the coveted pole starting position. Cale's velocity was nearly 9 miles per hour faster than Curtis Turner's pole-winning speed of just one year before. The secret of Cale's new-found speed was, of course, aerodynamics. And that did not bode well for Mopar drivers in general, and Plymouth drivers in particular, for the coming season. Though totally dominant during the 1967 season, Plymouth drivers were consigned to redesigned Satellite bodies that were every bit as boxy as their 1967 mounts. Benighted with barn-door-like vertical grille panels and formal sedan-style roofs, the new-for-1968 Satellite body styles were just about as aerodynamic as a brick. Dodge drivers had higher hopes for their new "fuselage" bodied Chargers, but those cars turned out to have hidden aerodynamic peccadillos of their own.

And so it was that 1968 turned out to be a "Total Performance" year for Ford and Mercury drivers on the tour. Yarborough translated his first-place starting berth at Daytona into a dominating performance during the 500. All told, his red and white Wood Brothers Cyclone led 86 of the event's 200 laps, including the all-important final one. Yarborough's closest competition that day was similarly named (but no relation) driver LeeRoy Yarbrough, who'd recently signed on to drive Junior

Floridian LeeRoy Yarbrough signed on to drive for Junior Johnson in 1968. He drove slippery Cyclones like this replica of his #26 team car. Though Mercurys that season featured vertical grille panels like their Torino siblings, a Cyclone still was better at air management due to its more angled hood line.

Nineteen sixty-eight became the Cale and LeeRoy show as both Mercury drivers were almost always running at the front of the pack. The secret of their success was the superior aerodynamics of their fastbacked Cyclone stock cars. By season's end, they scored 8 wins (including Cale's Daytona 500 triumph) and 27 top-five finishes between them. *Courtesy Ford Motor Company*

Johnson's #26 Mercury Cyclone. Cale and LeeRoy's one-two finish signaled the beginning of the Cale and LeeRoy show that dominated just about every superspeedway race during the 1968 season.

The Mercury duo reprised their Daytona finish at Atlanta in March with another one-two sprint across the line in the Atlanta 500. And that's just how Cale and LeeRoy finished in the Firecracker 400 at Daytona in July. LeeRoy finished out in front in the Dixie 500 at Atlanta when Cale crashed early in the event, and Cale's flashy #21 car crossed the stripe first at Darlington in the Southern 500 after LeeRoy's engine expired in that fabled event.

While Cale and LeeRoy were stealing the headlines in 1968, David Pearson was methodically racking up the points for the Holman & Moody team with a string of short-track wins and top-10 finishes. Pearson's #17 car finished first at Weaverville, Darlington (in the Rebel 400), Beltsville, Hampton, Macon, Bristol, Nashville, Columbia, Winston-Salem, Augusta, and Hickory. By season's end, his steady, methodical performance produced 16 wins, 36 top-5 finishes, and his second Grand National driving title (and the first won by the Holman & Moody team). All told, Ford and Mercury drivers scored wins at 27 of the 49 events on the 1968 schedule.

Mopar teams were none too happy with that result and resolved to trump Ford and Mercury's aero-advantage in the coming season.

The weapon that Chryco engineers devised to slay the Fomoco dragon was a slightly revised version of the Charger body style referred to as the Charger 500. In an attempt to cure the car's aerodynamic shortcomings, modifications had been made to the Charger's beak and roofline areas. More than a little overconfident that they'd solved their aero problems, Dodge execs took the unusual step of unveiling the new officially-approved variant at the fall 1968 race in Charlotte.

Unbeknownst to the Dodge boys, Ralph Moody was cooking up his very own aero-warrior in the super-secret back room of Holman & Moody's Charlotte Airport racing complex. Like his pentastar rivals, Moody spent time cleaning up the front end of the Ford Fairlane mule he was working on. To that end, he grafted on a new nose that extended the car's silhouette a full 6 inches and then finished it off with a flush-fitting grille and a labor-intensive front bumper rewelded to serve as an airfoil. Finishing touches were worked out by Ford stylists and wind tunnel engineers in time for Speedweeks 1969 at Daytona. And while they

Nineteen sixty-nine was David Pearson's second full season with Holman & Moody, and he spent that year behind the wheel of a blue and gold Talladega. As in 1968, Pearson proved to be the man to beat, and by season's end few had. As a result, he scored the second straight Holman & Moody national championship. *Courtesy Ford Motor Company*

were at it, the Fomoco aero-team cooked up a similarly configured aero-beak for the Mercury Cyclone line as well.

Fomoco execs called the new long-nosed Fords Talladegas after Bill France's still-under-construction palace of speed in Alabama. The Mercury version was dubbed the Cyclone Spoiler II. Ford plans called for the simultaneous introduction of its long-lusted-after corporate Hemi racing engine at Daytona. After having been denied track access for the Cammer 427 engine, Ford engineers had set to work on a cam-in-block version of the new (for 1968) "385" big-block engine that featured hemispherical combustion chambers cast in aluminum. When Dodge drivers showed up at Daytona with their sleek new Charger 500s, Ford teams were already in residence with a fleet of even slipperier Boss 429-powered Torino Talladegas.

With certain doom looming on the horizon, NASCAR stepped in at the last minute in an attempt to save the Mopar teams from total defeat. It seems that Big Bill France wasn't satisfied that Ford had actually built the requisite number of Boss 429 Mustangs in order to gain approval of the engine for use on the Grand National tour. As a result, France ordered that all the Boss engines be yanked out and replaced with last year's Tunnel Port 427s.

It was an interesting edict to be sure—and one that was more than a little suspect since, as history records, Dodge never came close to building the supposedly required 500 street-going versions of the new Charger 500 body style. But when the bombast and political maneuvering settled, Ford drivers were little hindered by France's last minute decision to "86" the Boss 429. David Pearson convincingly won one of the twin qualifying heats in his H&M-prepped (and Robert Yates-engined) Talladega. And in the race itself, LeeRoy Yarbrough translated a 19th-place starting position into his first Daytona 500 win.

Adding to the trounced Dodge drivers' depression as they filed out of the infield, tail

Fomoco's "A" team in 1969 consisted of (clockwise from right) LeeRoy Yarbrough, Donnie Allison, David Pearson, Cale Yarborough, and Richard Petty. They are all smiling here because they know they get to race Boss 429 hemi engines. *Courtesy Ford Motor Company*

between legs, was the fact that some of the slings and arrows of the unkind aerodynamic fate that had befallen them at Daytona had been hurled by former Plymouth partisan Richard Petty. Just before the season had begun, Petty had stunned the Mopar faithful by announcing his switch to cars of the Blue Oval persuasion. The reason for that unprecedented move was the exact same thing that had been the undoing of his former Dodge and Plymouth partners: aerodynamics. Simply put, when Petty got wind of Ralph Moody's swoopy new Talladega, he knew his chances of winning with a Plymouth in 1969 were just about nil. And so the King jumped ship to Ford's "Going Thing" for the 1969 season. It turned out to be a wise choice since, as he'd predicted, Plymouth drivers had more than a little trouble finding the way to the 1969 winner's circle (making it there just two times that season).

Though Ford teams were deprived of their "Blue Crescent" Hemis at Daytona, Big Bill relented in time for the Atlanta 500 in March and gave the all-new engine his official stamp of approval. Ford upped the aerodynamic stakes at that race even further with the simultaneous unveiling of the all-new Cyclone Spoiler II. Making matters even worse for Charger 500 drivers was the fact that the new long-nosed Mercs proved to be even faster than their Talladega siblings. That was due to the steeper angle that had been designed into the II's front sheet metal. Though only a matter of a few degrees, the extra angle added up to a 1- to 2-mile-per-hour advantage on the superspeedways.

Cale Yarborough put that asset to good use at Atlanta, where he led a dominating 308 of 334

Richard Petty broke Mopar lovers' hearts everywhere when he jumped ship to Ford in 1969. The reason for his change of heart is pictured here: a Ford Torino Talladega. King Richard knew his factory Plymouth ride was no match for the new Ford Aero-variant and decided, if you can't beat 'em, join 'em. He won 10 races in his long-nosed Ford that season.

total laps on his way to the winner's circle. It was a humbling defeat for the Dodge and Plymouth drivers, and one that, no doubt, sent them into an even deeper funk. For Fomoco drivers, it was the beginning of an incredibly successful season. LeeRoy Yarbrough scored a second Spoiler II win in May in the Rebel 400 and yet another two weeks later in the World 600. Cale upped the Mercury total to 4 with a win in the Motor State 500 at Michigan in June. Talladega drivers like David Pearson, Donnie Allison, and Richard Petty (and later in the season, LeeRoy Yarbrough) did their part in securing Blue Oval glory by winning big-track events at Rockingham, Dover, and Charlotte while also scoring 16 short-track triumphs. Petty's first road-course win (scored at Riverside) brought the Ford and Mercury win total to 30 for the season (26 Ford and 4 Mercury). David Pearson's Dick Hutcherson-led H&M team came out on top in the points chase, handing Holman & Moody their second straight Grand National title.

Fomoco's domination was so complete in 1969 that not even Dodge's desperate mid-season introduction of the radically winged Daytona (of which fewer than 500 were ever built—go figure) was not enough to derail the Blue Oval's winning ways. As things turned out, Dodge's gaudy new winged thing won but two superspeedway events. And one of those came at the inaugural Talladega event where most major teams (and every single Ford Talladega team) boycotted the event as the result of a safety dispute with Bill France.

Unfortunately, changing events behind the scenes in Dearborn would prove to be the undoing of Fomoco's aero-warriors during the 1970 season. Nineteen seventy was slated to be a model change year for Fomoco's intermediate lines. Longer, larger, and wider were the themes that held sway in styling studios circa 1970, and things were no different in Dearborn. Recognizing early that the new Fairlane and Montego body styles would not compare with their Talladega and Spoiler II predecessors

Talladegas and Spoiler IIs started out the 1969 season under 427 Tunnel Port power. When Bill France was satisfied that a sufficient number of Boss 429 Mustangs had been built to homologate that engine, Petty and other Fomoco teams shifted over to Boss '9 motorvation.

without help, Ford stylists like Larry Shinoda (who'd also designed the split window Corvette coupe, the Z-28, and the Boss 302) penned a radical new nose for both car lines that was designed to keep their aerodynamics apace with both their in-house and Brand X competition. In final form, Shinoda's design produced a front clip that rose in a smooth arc from pavement level to windshield base. Ford versions were called Torino King Cobras, while their Mercury siblings picked up the name Super Spoiler II.

Preliminary work on the two new aero-variants had progressed to the prototype stage by late 1969 when the rug was suddenly pulled from beneath the entire Ford racing program. The catalyst that brought that change was Lee Iacocca. It was Iacocca who got the nod from Henry Ford II when inveterate racer Bunkie Knudsen was booted from the Fomoco presidency. Unfortunately for Fomoco fans everywhere, as president, Iacocca

Ralph Moody and Larry Shinoda conspired to smooth out the aerodynamic flaws of the Cyclone and Torino car lines during the waning months of the 1968 season. Their solution was to graft on an all-new nose panel that both narrowed and lowered both cars' front body line. Fomoco execs called the new Mercury design the Cyclone Spoiler II. Cale Yarborough won the new line's first outing at the 1969 running of the Atlanta 500.

was diametrically opposed to continued corporate sponsorship of motorsports activities. One of his very first moves as Ford chief was to cut the corporate racing budget by a withering 75 percent across the board. As a result, the King Cobra project died before coming to fruition, and none of those sleek aero-warriors ever took a competitive lap on the Grand National tour. Worse yet, Talladega and Spoiler II teams that had been gearing up to campaign the new cars were forced at the last minute to abandon all of the R&D work done and hastily prepare their 1969 teams' cars.

The secret behind the Talladega's aerodynamic success was the special nose panel cooked up by Ralph Moody (with input from Larry Shinoda). It consisted of fenders that were extended 6 inches forward, a special header panel that tied the fenders together, a labor-intensive front bumper (cut and welded from a rear bumper), and a flush-fitting grille section. Donnie Allison drove the #27 car for legendary mechanic Banjo Matthews in 1969 and 1970.

LeeRoy Yarbrough had a phenomenal year for both Ford and Mercury in 1969. He kicked the season off by winning the Daytona 500 in a Talladega and then went on to score 6 more wins and 16 top-five finishes in both Talladegas and Spoiler IIs. He is shown here at speed in a long-nosed Mercury. *JDC Collection*

And, of course, Iacocca's budget axe also cut the flow of support dollars that flowed to factory teams to nearly a trickle. As a result, Fomoco's racing fortunes were mortally wounded before the first Boss 429 barked to life at Daytona, and the 1970 season was lost before it began.

A further impediment to Fomoco success in 1970 was the off-season introduction of yet another Chryco winged car. Stung by King Richard's 1969 departure for Ford, Plymouth engineers set out to seduce him back with an all-new winged Plymouth called the Superbird. That new aero-variant (coupled with several truck loads of money) was all it took to induce Petty to change his brand allegiance back to the Mayflower division.

And yet, with seemingly the whole world conspiring against them, Ford and Mercury drivers still managed to turn in a number of superspeedway wins during the 1970 season—with inadequate funding and absent radically pointed snouts or soaring rear deck wings.

The first of those wins almost came at the 1970 Daytona 500 where Cale Yarborough overcame a flock of winged cars in qualifying to turn in the pole-winning lap with his #21 Wood Brothers Spoiler II. Cale led the preliminary laps but was sidelined with engine failure early in the event. Holman & Moody driver David Pearson next picked up the Ford standard and late in the race appeared poised to win. Unfortunately, worn tires allowed Superbird driver Pete Hamilton to slip past just before the flag to take the win. Pearson did win at the spring Darlington race, and Donnie

Allison won both the World 600 and the Firecracker 400 in a Banjo Matthews-prepped Talladega. Cale Yarborough and LeeRoy Yarbrough scored superspeedway wins for Mercury at Michigan and Charlotte, and Cale's #21 car visited victory lane one more time on a big track (at Rockingham) before the year was over.

Though not a winning season overall for Ford and Mercury, the underfunded superspeedway wins scored with last year's race cars in 1970 helped secure final aero-war victory for Fomoco teams. When Bill France brought the aero-wars to an end in 1971 by restricting special aero-bodied cars to a maximum engine size of just 305 cubic

Nineteen seventy saw the introduction of an all-new Torino body line. Larry Shinoda penned an aeroversion of the new body style that featured a swoopy, Datsun Z-like front end that came to be called the King Cobra. Though the car showed promise, it died before it saw any racing time when Bunkie Knudsen was fired from Ford.

Here's a nose-to-nose comparison of the Talladega and the King Cobra. Early tests of the KC conducted by Holman & Moody showed that the new nose stuck to the ground like glue. Unfortunately, the car never got a chance to prove itself on the track.

Bud Moore bought two King Cobra prototypes from Ford and thus saved them from the crusher. Today, one of the two is in private hands and has been restored to Boss 429 trim. The other still belongs to Moore. Just imagine if Ford had gone forward with the KC race program.

Talladegas made appearances in a variety of racing venues. Benny Parsons, for example, raced in a Big T in both NASCAR and ARCA. His #98 Talladega finished up the 1969 season as the ARCA champ. *JDC Collection*

inches, Ford teams had won a grand total of 14 superspeedway races, with Mercury teams accounting for 8 additional mile or more wins. Winged Daytonas won just 6 superspeedway events during the same period and Superbirds just 7. Dodge Charger 500s accounted for but 1 big track win over the two-season affray. With a final tally of 22 superspeedway triumphs to Chryco's 14, it's clear that, pointy noses and soaring wings notwithstanding, Ford Talladegas and Mercury Spoiler IIs were the true champions of the factory-backed aero wars waged 1968 through 1970.

Ford went to war with the NASCAR sanctioning body over the legality of the SOHC Hemi engine. Ford lost that battle but ultimately won the war (the aero war that is) with the Boss 429 Hemi engine in 1969 and 1970.

The lights went out on Holman & Moody's fabled racing operation in 1972. Abandoned by Ford for a second time at the close of the 1971 season, the team built its last race car (pictured here) in 1972. The Boss 429-powered Torino was driven by Indy ace Bobby Unser at Riverside in 1973. Note the right-side fuel fill that was used on Riverside cars (where cars circled the track in a clockwise rather than counterclockwise fashion).

3

Ford's Lean Years
for Racing

Though few appreciated the fact at the time, 1971 was destined to be Ford's last season of factory-backed racing on the Grand National (soon to be called Winston Cup) circuit for the next 12 years. With funding cut to the bone by Iacocca's gutting of the racing budget, most teams that decided to continue campaigning cars of the Fomoco persuasion opted to cut back and only field cars at the most lucrative stops on the circuit.

And with Talladegas and Spoiler IIs relegated to the scrap heap by the vagaries of the official rules book, just about every Blue Oval operation switched over to 1969 Cyclone sheet metal, since it was felt those cars were more aerodynamic than any Fomoco intermediates that had been built since. Holman & Moody fielded

blue and gold Mercurys first for David Pearson that year, for example, and then (when Pearson left the team) red and gold Coca-Cola Cyclones for Bobby Allison. The Wood Brothers also switched to 1969 Cyclones for A. J. Foyt after Cale Yarborough's departure. That same move to Mercury was made by Banjo Matthews for his driver Donnie Allison. Long-time Ford partisan Junior Johnson jumped ship entirely for 1971 and elected to campaign Chevrolets for the first time since 1963.

In another interesting twist, most Ford teams ultimately abandoned their Boss engines in favor of tried and true 427 Tunnel Port engines when the restrictor plate regulation NASCAR had first introduced in 1970 began to choke the competitive life out of the "Blue Crescent" 429 engines

1971–1980

(because NASCAR's always-arbitrary rules book more severely restricted Hemi motors than conventional wedge engines).

As things turned out, 1971 was actually a pretty successful season for Holman & Moody. With Pearson still at the helm, the team won one of the twin 125s run every year before the Daytona 500 and finished fourth in the race. Pearson also turned in a win at Bristol before resigning from the team. Bobby Allison's first 1971 H&M effort produced a second (behind brother Donnie's Cyclone) at Talladega. The team then went on a superspeedway tear and won the World 600, the Mason Dixon 500, the Motor State 500, the Yankee 400, the Talladega 500, the Southern 500, and the National 500. Ralph Moody today recalls Allison's #12 Cyclone as his all-time favorite competition car (quite a statement when all of the cars that Moody drove or worked on are considered!). And on reflection, the stellar success enjoyed by the Holman & Moody effort that year was a fitting swan song for the team. For that's just what the season turned out to be as, when Ford removed all factory backing at the close of the year, the fabled Holman & Moody operation essentially ceased to exist.

The Wood Brothers' Stuart, Virginia, operation was also undergoing a transition away from factory sponsorship in 1971. Cale Yarborough had decided to leave the team at the close of the 1970 season to pursue his Indy car options, and A. J. Foyt had

When the sanctioning body effectively outlawed all special aerobody cars with a 5-liter limit in 1971, most Fomoco teams switched over to 1968/1969 Cyclone body styles. That proved to be a good move for the Holman & Moody team. When Bobby Allison stepped in for the departing David Pearson, he drove the #12 team car to wins in the World 600, the Motor State 400, the Yankee 400, the Talladega 500, the Southern 500, and the National 500. *JDC Collection*

Though not apparent at first, the 1970/1971 Cyclone body style turned out to be one of the most successful superspeedway body styles of all time. David Pearson and the Wood Brothers proved as much by winning just about every superspeedway race on the tour during the early 1970s. Pearson is shown here out in front at Daytona. *Daytona Racing Archives*

been signed to replace him. Foyt got off to a good start by finishing third in the Daytona 500 and then quickly found the way to victory lane at Ontario (an exact duplicate of the Indy circuit Foyt was master of) in the Miller High Life 500 in late February. Win number two for the team came in the Atlanta 500 in March. Donnie Allison hopped in Foyt's #21 "W" nose Mercury at Talladega and led older

brother Bobby across the stripe in the Winston 500 in what turned out to be the last win of the season for the Woods. All told, Mercury drivers scored 11 wins in 1971, while Ford drivers accounted for 4.

With Ford fully out of racing at the close of the 1971 season, it became increasingly challenging for the few remaining Fomoco teams to achieve victory. By the end of the decade, only a handful of diehard Ford partisans were still on a tour that by 1975 was almost exclusively a GM show.

Yet even during the darkest moments of that decade of despair, there were a few bright moments for Ford fans in the stands at Winston Cup stock car races. And more than a few of those moments were provided by the driver of a Wood Brothers Mercury.

A. J. Foyt, for example, gave a hint of the superspeedway domination the Woods' team would display in the mid-1970s when he convincingly won the 1972 Daytona 500 in a team Cyclone. One of the secrets behind Foyt's dominating success that day (he started second and lead the last 300 miles of the event) was the overlooked aerodynamic prowess of the 1970 Mercury Cyclone body. Foyt used that advantage to win at Ontario just three weeks after Daytona. Though only allowed a vestigial 1 1/2-inch rear deck spoiler by the official rules book, the 1970/1971 Cyclone body was actually one of the best superspeedway chassis to ever wear a coat of racing livery.

Proof of that fact is the incredible string of superspeedway wins scored by David Pearson after he signed on to drive the car in April of 1972. Pearson, who was soon to earn his now famous "Silver Fox" nickname, won the first time out in a #21 Mercury at Darlington in the Rebel 400. A win in the Talladega 500 came next and was quickly followed by victory lane visits at Michigan (twice), Daytona, and Dover.

Independent drivers like Cecil Gordon also selected the slippery 1971 Cyclone body style as the basis for their racing efforts. This particular example has been restored by Davidson, North Carolina's Alex Beam.

Nineteen seventy-three found the Silver Fox once again behind the wheel of a 1970/1971 Cyclone, and once again that combination proved to be a winner. The team (and Pearson) had cut back its schedule to just 18 events that year, and all of those were super-speedway contests. Nonetheless, Pearson and the Woods won 11 of the events they entered. In that number were triumphs at Rockingham, Atlanta, Darlington, Martinsville, Talladega, Dover, and Daytona.

Though long associated with General Motors cars, Darrell Waltrip actually cut his NASCAR teeth on Mercury race cars. Ol' DW's first Grand National outings came in this #95 1971 Mercury which he's recently restored to race-ready condition. *Craft Collection*

While Pearson was wowing the crowds (and the competition) with high-speed Mercury victories, cars carrying Ford's Blue Oval badge never got closer to victory lane than pit road the whole season. Even so, canny mechanic Bud Moore was laying the ground work for future Ford greatness that very season. You see, though Pearson was still mopping up on the high banks with the help of big block (Boss 429) power, the sanctioning body was taking an increasingly jaundiced view of 7-liter powerplants. As a result, the ever-changing rules book was making life increasingly difficult for big-block-based teams. Innovative types like Moore (just back from the Trans-Am wars where small blocks reigned) quickly discovered that smaller displacement engines were far less regulated by the infernal rules book than mountain motors were.

As a result, Moore set out to build a competitive Winston Cup car that was motorvated by a racing version of the 351 Cleveland small block first introduced in 1970. Beginning in 1973, Moore built cars for Bobby Isaac, Darrell Waltrip, Donnie

Allison, George Follmer, Buddy Baker, and Bobby Allison, and every one of those #15 Fords was pulled around the race track by a small-block Ford engine. It was one of Moore's Cleveland-powered cars, in fact, that ended Ford's three-year drought with a win in the 1975 Winston 500 at Talladega. Proving that win was no fluke, Baker flat-footed his way to a second 1975 Talladega win in the Talladega 500. Baker also scored firsts at Atlanta and Ontario that season, and by year's end, more than a few teams were reevaluating their reliance on big-block engines.

The Wood Brothers continued their super-speedway mastery throughout the mid-1970s, winning 7 events in 1974, 3 in 1975, and 10 in 1976. The most famous of those wins is, without doubt, the dramatic Daytona 500 triumph Pearson scored in 1976. When the 1970/1971 Cyclone body style that had served the team so well was rendered obsolete by the rules book in 1974, the Wood Brothers team switched to the newer (and not nearly as aerodynamic) Mercury Montego body style. The team had also elected to follow Bud Moore's (and the rule book's) lead at about the same time and rely on small block motorvation.

Pearson had used the 600 or so ponies cranked out by his Merc's Cleveland to qualify seventh for the race. Nonetheless, he had the boxy Montego out in front of the pack by lap 5 of the event. By lap 155, the race had become a two-car affair to be decided between Pearson and his arch-rival at the time, Richard Petty. They traded the lead back and forth for the next 50 circuits. Petty had claimed the lead with 13 laps to go, and Pearson stayed glued to his bumper. As the speeding duo approached Turn 3 on the last lap of the event, Pearson made his move. He'd completed his pass but then drifted up

This is the view that most of the NASCAR field had of David Pearson's 1971 Cyclone during the early 1970s. The car's native speed coupled with Pearson's crafty driving won more than a few superspeedway events.

Bud Moore returned to the NASCAR fold (from the Trans-Am ranks) in 1972. The first cars he built that season were for team driver Bobby Isaac. One of those gape-nosed Torinos has recently been restored by North Carolina's Kim Haynes.

the banking in a fight for traction. Petty took the opportunity and attempted to slip back by on the low side of the track. Petty too lost his purchase on the asphalt, and there was contact between the two cars that sent both spinning wildly into the tri-oval at nearly 180 miles per hour. Pearson's car hit the outside wall hard in the front, crazily distorting the sheet metal back over the engine and chassis. Petty's STP Dodge was similarly wounded, and when both cars came to rest, for a moment it appeared that neither would win the race. Fortunately, Pearson had the presence of mind to keep the clutch down and the revs up while spinning out of control (they didn't call him the Silver Fox for nothing), and so he was able to coax his heavily damaged Mercury across the line ahead of both Petty and the onrushing field. It was a spectacular win, and one that was broadcast live to a country full of television viewers who were perched on the edge of their seats.

The car that Pearson limped into victory lane with was a reflection of the technology that dominated the Winston Cup scene from the mid-1970s until nearly the present. Unlike the unit-body-based "half-chassis" cars that had been introduced by Bud Moore during the boycott 1966 season, by 1974 Fomoco teams had shifted to full-perimeter chassis that featured fully fabricated front and rear snouts. For a time in the middle of the decade the rules book still required the use of stock chassis side rails, but even that wink at stock appearance had gone by the wayside by the end of the decade. Suspension components under Pearson's Merc were a mix and match of Ford and GM components. A familiar Ford 9-inch live axle was mounted aft, but it was reined in by a pair of Chevy-truck-derived trailing arms and a set of screw-jack-adjusted coils. Four shocks were employed to keep jounce in check. Fabricated "A" arms were used at the bow, and they captured another set of screw-jack-adjusted springs. A 1965 Galaxie steering setup was used to dial in directional changes. Two more pairs of shocks and a wrist-thick, through-the-chassis sway bar rounded out the front suspension.

Disc brakes had become increasingly popular in the garage area ever since Mark Donohue had scored the first NASCAR disc brake win in 1972 (in a Rambler of all things!), so Pearson's Mercury that day was stopped by a stout set of aftermarket rotors and calipers.

The rest of his Mercury's chassis consisted of the jungle gym's worth of tubing that the NASCAR rules book had mandated by the late 1960s. Four bars were mounted per side, and the basic cage had expanded to eight major mounting points. A Petty Bar and a handful of diagonal braces had also become part of the safety recipe by the time of Pearson's last lap shunt, and they no doubt helped to keep him safe and sound and able to drive across the stripe first.

The whole package rolled on 15x9.5 aftermarket rims and treadless Goodyear bias-ply tires— treaded racing rubber having gone to its eternal reward in 1971. Other significant bits and pieces of the package included a safety fuel cell (first required by the rules in 1967), a foam-based fire system (first required in 1971), and, of course, a full-race Cleveland small block cooked up by Leonard Wood.

Ford performance, on the other hand, was not exactly noteworthy during the mid-1970s. Buddy Baker won yet another Talladega race in a Bud Moore Ford in 1976, but that would be the only entry in the Ford win column until 1978.

When NASCAR determined that speeds were too fast for safety (sound familiar?) in the early 1970s, the sanctioning body set out to eliminate the use of 7-liter engines. Bud Moore countered that assault with a series of small-block (Cleveland) based race cars. Buddy Baker (pictured here during a pit stop) and others drove for Moore during the 1970s and scored a handful of superspeedway wins despite their pioneering use of 5.8-liter powerplants. *Daytona Racing Archives*

Exiting the NASCAR scene in 1972 was the famed Holman & Moody race car-building team. It is perhaps fitting that their last car built (pictured at beginning of chapter) rolled out of the Charlotte shop under the power of a full-boogie Boss 429 engine. Note the late-style ram air system the car came equipped with.

Bobby Allison was an on again/off again Fomoco driver throughout his career. The late 1970s were an "on again" period for Allison, when he drove for Bud Moore's Spartanburg, South Carolina-based team. Though bulky, the T-Birds he drove did enjoy superspeedway success. *Daytona Racing Archives*

Pearson continued to turn in Mercury wins for the Woods, but at a much reduced pace compared to the early 1970s. He scored two wins in 1977, (including the Southern 500), and four in 1978 (including the Firecracker 400) before ending his association with the Wood Brothers on an unhappy note at Darlington (after he left the pits with no lug nuts in place). Neil Bonnett took Pearson's place during the next two seasons and scored five more Mercury wins (including superspeedway wins at Dover, Daytona, Atlanta, Pocono, and Talladega). The last of those wins came at the 1979 Talladega 500 where Bonnett started second on the field and then went on to lead Cale Yarborough's Junior Johnson Chevrolet across the stripe. It is more than a little ironic that Bonnett's win that day, just ahead of former Mercury star Yarborough, should have been the very last to be recorded by a Mercury driver in current NASCAR history. Though some teams have built and tested Lincoln Mercury-based race cars in the nearly two decades since that day in 1979 (Cale Yarborough's team experimented with a Cougar in

the 1980s, and Michael Kranefuss' team tested a Lincoln Mark VIII in the 1990s), no other Mercury drivers have visited victory lane. Indeed, there are no Mercury-based teams in competition on the circuit today and likely will not be in the near future.

As Mercury's star was in decline in the late 1970s, Ford drivers (at least one Ford team, that is) were experiencing some measure of success. When Buddy Baker and Bud Moore parted ways in 1978, Bobby Allison signed on to campaign the team's block-long #15 Thunderbirds. Allison had driven Cam 2 Mercurys for Roger Penske the season before, but that operation had folded its tent at the end of the year (which resulted in the sale of several team Mercurys to a young Georgia fellow named Bill Elliott who then used them in some of his earliest NASCAR appearances). So the senior member of the Alabama gang was amenable to Moore's offer when it came during the off season. It proved to be a very successful pairing.

Evidence of that fact is provided by the 1979 win that Allison turned in for the team at the all

Fomoco racers in the 1970s relied on the junkyard (rather than Dearborn) for racing parts. This Wood Brothers 351 C engine, for example, was built around a scavenged Australian block and regular production heads. Fortunately, Ford's interest in factory-backed racing would awaken from its dormancy in the early 1980s.

important Daytona 500. Though Bobby had only qualified his boxy T-Bird 33rd that day, by the end of the event he placed the car's block-long hood out in front of the Chevrolet hordes. It was the first win in the "Super Bowl of Stock Car Racing" for both Moore and Allison. Also of note that day was the 8th-place finish turned in by that young red-headed fellow named Elliott in one of Allison's old Cam 2 Mercurys. It was Elliott's first top-10 finish on the Winston Cup tour, but it was to be far from his last. Allison finished 2nd at Rockingham in the #15 car before snagging his second win of the year at Atlanta. Win number three came at Dover in

Red Farmer was a charter member of the Alabama gang. As a result, the superspeedway at Talladega, Alabama, was one of his favorite tracks. The 1973 Torino he raced there is fairly typical of Ford stockers of the 1970s: aerodynamically impaired and down on power. Better days were ahead for Ford racers in the 1980s. *Mike Slade*

Here in a picture is Fomoco's problem during the late 1970s: aerodynamics or, specifically, the lack thereof. Where once slippery aero-warriors like the Torino Talladegas had knifed through the air, by the late seventies, the Ford styling studio had regressed to parasitic-drag producing grilles like the one on this Wood Brothers Mercury. Lack of factory support, of course, didn't help put Fomoco cars in the winner's circle either. *JDC Collection*

September, and Allison went on to win at Charlotte and in the season finale at Ontario. Allison stayed with Moore's Spartanburg-based operation through the 1980 season, and during that time he scored nine more wins for the team.

Things were looking bleak for Ford teams and fans as the 1980s dawned. Save for the efforts of the Wood Brothers and Bud Moore teams, the 1970s would have been a winless decade for Blue Oval partisans. Chevrolet and its GM clone divisions mopped up in the Winston Cup division during Fomoco's absence. Though not openly funding Bow Tie-based racing programs, as in the early 1960s during the AMA ban, Chevrolet engineers had left the back door of the high-performance division open throughout the 1970s. Ford, with Iacocca at the helm, instead had focused its attention on such stunning concepts as the Maverick

Grabber, the Mustang II, and the Pinto wagon. The end result was a lost decade for Ford racers—except of course—for those like Junior Johnson who had jumped ship to Chevrolet.

Though it's hard to imagine now, by 1979 it was becoming nearly impossible for even diehard Fomoco teams like Moore's, the Woods', and Junie Donlavey's to press on, due to the increasing rarity of usable Cleveland engine components. With that engine out of production and junkyard parts dwindling, Ford teams were forced to prowl Australian junkyards in search of race-worthy blocks and heads. Fortunately, Iacocca was soon to depart Ford for Chrysler (where he fathered the boring "K" car line), and with his departure, racier lights in the Ford executive class (in whose number was no lesser personage than Edsel Ford II) began to move the corporate tiller back towards motorsports competition.

Left
Though the late 1970s were the doldrums for Ford fans on the NASCAR tour, they did produce future reasons for enthusiasm—like Bill Elliott. The yet-to-be-Awesome Bill got his NASCAR start at the wheel of dirigible-class Mercurys like this one during the latter part of the leisure-suit decade. *JDC Collection*

Below
Ford and Mercury race cars had become all angular and bulky by the late 1970s, as can be seen in this trio of Mercurys and Fords. Bill Elliott is piloting the #9 Mercury, Bobby Allison the #15 Thunderbird, and David Pearson the #21 Merc. Rounded edges and regular returns to the winner's circle were still several years in the future. *JDC Collection*

Davey Allison, pictured here at speed, had his racing career tragically cut short by a freak helicopter accident in 1993. He was a bright, personable young man whose future was unlimited. In that regard he had more than a little in common with Fireball Roberts, Joe Weatherly, and Alan Kulwicki—all Ford racers who came to an untimely end.

4

Ford Racing
Roars Back to Life

Nineteen eighty-one was the year that NASCAR downsized its chassis from 116-inch wheelbases to the 110-inch standard that is currently in effect. The new smaller Thunderbirds produced by this rules change were both nimbler and more competitive than the Frank Cannon-mobile full-sized cars they replaced. And that fact is borne out by the nine wins that teams fielding those cars scored in 1981 and 1982.

Bill Elliott scored his very first Winston Cup win in the Winston Western 500 at Riverside (November 1983) at the helm of a downsized 1982 Thunderbird. Bud Moore fielded winning Fords for Benny Parsons and Dale Earnhardt (that's right, Ford fans, "Ironhead" once drove Fords!), and Neil Bonnett also added to the Wood Brothers Fomoco win total.

Though Ford teams once again began to taste stock car victory in the early 1980s, it wasn't until 1983 that the changes necessary to return Ford to a position of dominance began to occur. The most important event in that number was the decision taken in the Ford styling studio to move away from sharp edges and towards rounded, flowing surfaces. That change of focus led directly to the totally restyled Thunderbird line that took a bow in 1983. As in 1968, the power of superior Ford aerodynamics began to be felt at tracks all across the circuit.

At the same time those curvy new creations began to roll off of the assembly line, "Glass House" execs gave the green light for the production of a new run of 351 Cleveland-based engines and high-performance parts.

1981-Present

NASCAR rules dictated downsizing in 1981 from the traditional 116-inch wheelbase of years gone by to a new shorter 110-inch wheelbase. That change brought about the introduction of Fox chassis-based Thunderbird race cars. The Wood Brothers carried their familiar red-and-white #21 livery into the 1980s with Neil Bonnett as their driver (seen here during a pit stop at the North Carolina Motor Speedway). *Mike Slade*

More encouraging yet was the fact that, for the first time in nearly two decades, the corporation began to offer a catalog full of over-the-counter high-performance aftermarket parts. While Ford's revived interest in factory high performance wasn't exactly a rebirth of the halcyon Total Performance days of yore, it was a dramatic change in focus compared to Iacocca's days of thrifty fuel sippers.

Ford drivers Dale Earnhardt, Buddy Baker, Bill Elliott, and Ricky Rudd scored eight wins in 1983 and 1984. Their on-track success did not go unnoticed. Soon (and for the first time since the late

1960s) new Ford teams began to form. One of the first GM-based teams to switch back to Ford was Cale Yarborough's Ranier Lundy-backed operation. Yarborough's name already appeared many times in both the Ford and Mercury win columns, so his return to the Blue Oval was something of a homecoming. So, too, was the reunion of Ranier Lundy chief mechanic Robert Yates with the new and improved Cleveland-based engines that were slated to power Cale's #28 Hardee's car. As history records, a much younger Yates was once a Holman & Moody line mechanic and had, in fact, built the Tunnel Port engine that had powered LeeRoy Yarbrough to victory lane in the 1969 Daytona 500. Yates-built Ford engines were soon making themselves right at home in victory lane again.

Nineteen eighty-five was the year that Ford teams signaled that they intended to dominate the Winston Cup series in the same way their spiritual predecessors had in the 1960s. Bill Elliott fired the first volley of that assault in the Daytona 500 where, with the help of a high-revving small block built by brother Ernie, Awesome Bill swept qualifying with a record shatter-ing 205.265-mile-per-hour qualifying lap. Elliott used his car's speed advantage during the race itself to wear down the competition with a frantic 192-mile-per-hour average pace. He went on to lead 136 of the 200 laps that made up the event, including the all important final circuit. Elliott also turned in wins in Atlanta, Darlington, Talladega, Dover, Michigan, and Pocono. Along the way he captured the Winston Million and became the media's darling. His total tally of wins came in at 11, and his earnings topped two million simoleons. Interestingly, though Chevy

Dale Earnhardt's Fox-based Thunderbird wasn't any more aerodynamic than the rest of the breed, but that didn't stop him from winning the 1982 Rebel 500 in one such car.

Bill Elliott took his first checkered flag from a cockpit just like this one in 1983. The Melling team was less than generously funded in those days. Evidence of that fact was found in this particular car during its restoration by Alex Beam: the driver's seat was found to be made up from a "liberated" Georgia State road sign.

Previous pages
Fomoco racing fortunes took a dramatic turn for the better in 1983. The reason for that change was that old reliable physical force—aerodynamics. After years of building sharply-creased, vertically-paneled cars, Ford stylists went ovoid in 1983. The end result was super-slippery race cars like this Cale Yarborough Hardee's T-Bird. *Mike Slade*

rival Darrell Waltrip won but 3 Winston Cup events that year, the always-confusing NASCAR points system made him the Winston Cup champ. Superspeedway wins by Cale Yarborough and Ricky Rudd brought Ford's win total to 15 for the season.

As fast as Ford's first generation of rounded-off Thunderbirds had proven to be, even faster velocities were in the offing for 1987. That heightened performance came as a result of yet another styling change that refined the Thunderbird silhouette to what many still consider to be its most

aerodynamic form. Of greatest significance were the changes made to the car's bustle that both raised and widened the body line. The nose received refinement, too, and in race trim the body line rose from the racing tarmac to the "A" pillars in a smooth arc. When coupled with the ever-increasing levels of horsepower that were being found by Ernie Elliott and Robert Yates, speeds well in excess of the double ton were just a tickle of the throttle away.

Bill Elliott literally drove that point home during qualifying at Daytona in February when he scorched the track with a 210.364-mile-per-hour lap. Starting from the pole, he went on to humble the rest of the field by leading over half of the event and taking the checkered flag. Even the new add-on noses and bubble-back rear windows that the GM intermediates had sprouted to break the Thunderbird's stride proved unable to slow Awesome Bill's pace. Ford wins came quickly at other superspeedways on the tour. Ricky Rudd won at Atlanta and later at Dover. Rookie Davey Allison scored his first NASCAR win at Talladega behind the wheel of a Robert Yates missile (after Elliott eclipsed the field by qualifying with a record-setting 212.809-mile-per-hour pace lap). Allison won again at Dover. Elliott claimed the Talladega 500 (where the summer heat slowed his qualifying speeds to a tepid 203.827 miles per hour). He won at Michigan, then again at Charlotte, once more at Rockingham, and for the final time at Atlanta.

All told, Ford drivers won 11 times in 1987. And that was enough to persuade the sanctioning body that something would have to be done about, ahem, safety (yeah, that's it; it's just plain unsafe for Ford teams to win so many races). Reaching into the tech inspector's golden oldie bag of tricks, NASCAR announced that the 1988 season would see the reintroduction of the dreaded restrictor plate. Even so, Bill Elliott still visited victory lane a total of 6 times and scored 9 other top-10 finishes on his way to the 1988 Winston Cup driving title. It was the first such win for a Fomoco driver since David Pearson's 1969 triumph for Holman & Moody and had certainly been a long time in coming.

With those horsepower-robbing plates in place, the 650-plus horsepower that folks like Yates could coax from just 5.7 liters of Ford small block were trimmed a substantial 200-odd ponies. And that was enough to keep Ford teams out of victory lane at more than a few events. And, oh yes, the resulting bunch-up of cars that often led to huge chain reaction wrecks was much safer indeed than when Ford drivers were lapping the field twice at Talladega under green flag conditions. Unfortunately, the dreaded restrictor plates first rolled out in 1988 are still with us today.

Early 1980s' T-Birds sported formal, chopped-off rooflines that were great at creating drag. No restrictor plates were needed in those days to keep speeds in check.

Davey Allison was a rookie driver during the 1987 season. But his lack of experience did not keep his Robert Yates-prepped Ford out of victory lane. Young Davey became the first rookie to win more than one race in his freshman season since Dick Hutcherson earned that same honor as a Holman & Moody driver in 1965.

When Ford decided to get reinvolved in racing, the corporation began to crank out sorely needed high performance engine parts. By 1983, it was again possible to build a full race 351 C-evolved engine from all new components. No more junkyard shopping was required.

Even in the face of NASCAR's best efforts to slow the Ford contingent down, more and more teams began to make the switch to Thunderbird-based teams. Whereas in the 1970s, there might have been just one or two Fomoco-based teams on any given starting grid, as the 1990s dawned, oftentimes more than half of the field was comprised of cars bearing the Blue Oval. The Thunderbird's slippery shape, coupled with an increasingly robust behind-the-scenes support from Ford, made it a no-brainer for many teams to switch to Ford.

The 1990s have been marked by continued Ford success. Nineteen ninety-two saw Ford drivers visit victory lane no fewer than 16 times. In that number was a memorable (and sentimental)

When Ford stylists revised the Thunderbird body style in 1987, part of the program was to raise the car's rear deck lid several inches. The line's front silhouette was also freshened up that year and made more aerodynamic. The end result was the fastest NASCAR stock car of all time. Bill Elliott's 212.809-mile-per-hour qualifying lap at Talladega in 1987 will likely never be topped.

win scored by Davey Allison in the Daytona 500. Young Davey, whose life would be taken in a tragic flying accident less than a year later, started the race sixth that day (behind fellow Ford driver Sterling Marlin who had a 192.213 qualifying lap in the Junior Johnson Ford). Though he didn't make it to the front of the pack until lap 56 of the event, Davey led the last 29 circuits convincingly to take the win. Bill Elliott (running for Junior Johnson) and Davey Allison both wound up with 5 wins each that season, but it was the doggedly independent Alan Kulwicki who came up with the points to secure the Winston Cup driving title. Kulwicki's formula for success included a handful of Thunderbirds (sometimes called "Underbirds" due to his

Bill Elliott took in the very rapidly passing scenery from this perch in 1987 and 1988. Interestingly, just beneath the floorboards lurked most of the same suspension components that cars like Fred Lorenzen's 1965 Galaxie had rolled on.

Into the 1990s, the Wood Brothers are still fielding competitive cars of the Ford persuasion. Though no longer painted in the Wood's traditional red-and-white livery, the race cars that they build in Stuart, Virginia, still carry the familiar #21 racing number. Michael Waltrip is the most recent of the team drivers.

The Wood Brothers continued their winning ways in the 1980s with drivers like Neil Bonnett and Kyle Petty. Here the team services young Kyle's T-Bird on the way to a win in the 1987 World 600.

Ricky Rudd proved to be one of the most consistent Ford drivers in the 1990s, winning at least one Winston Cup race every season for nearly a decade. His Tide-backed T-Birds were both flashy and fast.

independent status), 2 wins (at Pocono and Bristol), 17 top-10 finishes, and lots of hard work. The end result was $2,332,561 in winnings and the second Winston Cup driving title secured by a Fomoco driver in the modern era.

Ford drivers scored a modern-era high of 20 season victories in 1994, and Jack Roush driver Mark Martin came within a hair's breadth of winning the third Winston Cup title of the modern era. The 1990's iterations of the Thunderbird chassis are much the same as the early 1980's Fords they succeeded. All are built around a purpose-built tubular-steel chassis that has never been closer to a UAW assembly line than a North Carolina fabricator's shop. A 9-inch solid-axle differential forms the focus of the car's rear suspension, and its movements are checked by a pair of Chevy-derived trailing arms, a cross-chassis panhard rod, and a pair of screw-jack-adjusted coil springs. One gas-charged shock per wheel keeps

Bud Moore's #15 Fords continue to make regular appearances well into the 1990s. Recent drivers of Moore-prepped T-Birds include Ricky Rudd, Geoff Bodine, and Dick Trickle. In 1998, the Spartanburg native celebrated his 50th year in stock car racing.

oscillations in check. At the box, most teams have now switched to Chevrolet Camaro-derived front-steer (steering box ahead of center line) suspension components that allow greater flexibility in header and oil pan configuration. Another pair of gas shocks and a cross-chassis spline-ended sway bar complete the front suspension. Special manhole-cover-sized discs and six-piston caliper brakes are mounted at all four corners, and they act on 15x9.5 rims that roll on special Goodyear Eagle radial tires.

Power for a modern Winston Cup T-Bird is churned out by an alloy-headed evolution of the 351 Cleveland engine. A single 850-cubic-feet-per-minute-based Holley carb sends combustibles to the engine's poly-valved combustion chambers

Alan Kulwicki was the epitome of the rugged individualist. Though offered several "big time" rides with long-established teams, he chose to go his own independent way. That course led him to the 1992 Winston Cup championship.

through a highly massaged single-plane aluminum intake. A cobby roller cam incapable of idling below 2,000-odd rpm governs timing events, and a high-energy ignition (HEI) system fires the engine to life. A low-parasitic-drag dry-sump system carrying 20-plus quarts of synthetic provide the engine's life blood, and handmade large-diameter stainless headers send hydrocarbons out towards the ozone layer. In peak Robert Yates tune, a modern high-compression Ford engine can be relied on for in excess of 725 ponies. Absent the restrictor plates that NASCAR still mandates at Daytona and Talladega (among other superspeedways), it's likely that a Winston Cup "Bird" would be circling the track in excess of 220 miles per hour.

Not happy with that fact, the sanctioning body has low compression ratio 9:1 engines waiting in the wings and is moving in the direction of cutting speed on the track by reducing the allowable squeeze in each engine's combustion chambers.

In addition to these changes, Ford itself has discontinued the Thunderbird line for 1998. It appears that 1997 will be the last year for a Thunderbird on the circuit (at least until the new Thunderbird line makes a bow in 1999 or the year 2000). However, for the 1998 season Ford introduced a Winston Cup version of the four-door Taurus economy car that is showing promise.

Evidence of its potential was provided by Rusty Wallace's Taurus triumph in the 1998 running of the Bud shootout—the new body style's first competitive outing. More Taurus victories

Ned Jarrett won the 1965 Grand National driving championship in a Holman & Moody-built Ford Galaxie. In the 1990s, his son Dale has shown the same flair for piloting a race car that his dad had in the 1960s. The younger Jarrett will without doubt duplicate his dad's NASCAR driving title in the very near future.

Mark Martin's #6 Taurus quickly established itself as a championship contender during its first season of Winston Cup Competition.

were quick to follow. As this is being written, the 1998 season is winding down to a head-to-head battle between Mark Martin's #7 Taurus and Jeff Gordon's Monte Carlo. Though few predicted that Ford's new body style would show championship potential right out of the blocks, that's how things have turned out. But then again, Ford does, after all, stand for First On Race Day, doesn't it?!

Since the very first strictly stock race in Charlotte in 1949, Ford drivers have visited Grand National and Winston Cup victory lanes more times than any other manufacturer (as of press time that figure is 479). And that's a winning tradition that's likely to continue well into the next century.

Ricky Rudd's Taurus was one of the most colorful cars on the tour in 1998 regardless of brand affiliation.

Appendix

First Model Wins

First	Laps/ Track Length	Place	Date	Driver/ Team/Car
Lincoln Win	150 laps 3/4-mile dirt	Charlotte, North Carolina	6/19/49	Jim Roper Mecklenburg Mtrs 1949 Lincoln
Mercury Win	200 laps 1/2-mile dirt	Vernon, New York	6/18/50	Bill Blair 1950 Mercury
Ford Win	200 laps 1/2-mile dirt	Dayton, Ohio	6/25/50	Jimmy Florian Euclid Motors 1950 Ford
Thunderbird Win	110 laps .9-mile dirt	Hillsborough, North Carolina	3/1/59	Curtis Turner Holman & Moody 1959 T-Bird

First Championship

First	Year	Driver	Team/Car	Wins	Top 5 Finishes	Purse
Ford Grand National Championship	1965	Ned Jarrett	Bondy Long Galaxie	13	42	$93,624.40

First Race Wins

First	Year	Driver	Team/Car	Avg. Speed (mph)	Purse
Ford Daytona 500 win	1963	Tiny Lund	Wood Brothers Galaxie	151.566	$24,550
Mercury Daytona 500 win	1968	Cale Yarborough	Wood Brothers Cyclone	143.251	$47,250
Ford Southern 500 win	1956	Curtis Turner	Schwam Motors Fairlane	95.067	$11,750
Mercury Southern 500 win	1966	Darel Dieringer	Bud Moore Engineering Comet	114.830	$20,900
Ford World 600 win	1962	Nelson Stacy	Holman & Moody Galaxie	125.552	$25,505
Mercury World 600 win	1969	LeeRoy Yarbrough	Junior Johnson Cyclone Spoiler II	134.361	$29,325
Ford Talladega 500 win	1975	Buddy Baker	Bud Moore Engineering Torino	144.948	$28,725
Mercury Talladega 500 win	1971	Donnie Allison	Wood Brothers Cyclone Spoiler	147.419	$31,140

First Pole Positions

First	Year	Driver	Team/Car	Speed (mph)
Ford Daytona 500 Pole	1985	Bill Elliott	Melling Racing Thunderbird	205.114
Mercury Daytona 500 Pole	1965	Darel Dieringer	Bud Moore Engineering Marauder	171.151
Ford Talladega 500 Pole	1985	Bill Elliott	Melling Racing Thunderbird	209.398
Mercury Talladega 500 Pole	1971	Donnie Allison	Wood Brothers Cyclone Spoiler	185.869

Fomoco Driver Records

Most Ford Grand National/Winston Cup Championships: David Pearson—2 (1968 and 1969)

Most Ford Grand National/Winston Cup wins: Ned Jarrett—43

Most Mercury Grand National/Winston Cup wins: David Pearson—44

Fomoco Team Records

Most Ford Grand National/Winston Cup Championships: Holman & Moody—2 (1968 and 1969)

Most Ford Grand National/Winston Cup wins: Holman & Moody—84

Most Mercury Grand National/Winston Cup wins: Wood Brothers—63

Other Fomoco Records*

*Ford Manufacturer's Points Championships:
11—1956, 1957, 1963, 1964, 1965, 1967, 1968, 1969, 1992, 1994, and 1997

*Most Grand National/Winston Cup wins of any manufacturer: Ford—479

*Most Grand National/Winston Cup wins scored in a single season: Ford—48 (1965)

*Winningest Ford car line: Thunderbird—173 (1959–1997)

*Winningest Mercury car line: Montego—77 wins (1968–1980)

*Winningest Ford body style: 1989–1997 Thunderbird—115 wins

*Winningest Mercury body style: 1968–1969 Cyclone Spoiler—24 wins (1968–1971)

*Greatest number of top-five finishes in a single season:
Ned Jarrett (1965 Bondy Long Galaxie) and David Pearson (1969
Holman & Moody Torino Talladega) are tied with 42

*Largest margin of victory: Ned Jarrett (1965 Southern 500 in Bondy Long Galaxie)—14 laps

*NASCAR's first Most Popular Driver: Ford driver Curtis Turner (1956)

Index